Cosmic Butterflies

The Colorful Mysteries of Planetary Nebulae

At the end of a star's life, it wraps itself in a cocoon by spilling out gas and dust. Sometime later, a butterfly-like nebula emerges from the cocoon and develops into a planetary nebula. Planetary nebulae are among the most beautiful of the celestial objects imaged by the *Hubble Space Telescope*. Their structures, like bubbles floating in the void, are complemented by a kaleidoscope of color emitted by glowing gases. Delicate, lacelike streamers of gas add to their complexity. The production of a planetary nebula is a milestone in the life of a star, an event that foretells the doom of the star before it disappears into oblivion. In this book, Sun Kwok tells the story of the discovery process of the creation of planetary nebulae and of the future of the Sun.

SUN KWOK is a leading world expert on planetary nebulae. He is Professor of Astronomy at the University of Calgary and a Canada Council Killam Fellow. He serves as chairman of the planetary nebulae working group of the International Astronomical Union. A Canadian citizen, he is committed to the development of astronomy in Asia.

Sun Kwok

Cosmic Butterflies

The Colorful Mysteries of Planetary Nebulae

PUBLISHED BY THE PRESS SYNDICATE OF THE UNIVERSITY OF CAMBRIDGE
The Pitt Building, Trumpington Street, Cambridge, United Kingdom

CAMBRIDGE UNIVERSITY PRESS
The Edinburgh Building, Cambridge CB2 2RU, UK
40 West 20th Street, New York, NY 10011-4211, USA
10 Stamford Road, Oakleigh, VIC 3166, Australia
Ruiz de Alarcón 13, 28014 Madrid, Spain
Dock House, The Waterfront, Cape Town 8001, South Africa

http://www.cambridge.org

First published 2001

Printed in the United Kingdom at the University Press, Cambridge

Typeface Trump Mediaeval 11/16pt. *System* QuarkXpress® [HMCL]

A catalogue record for this book is available from the British Library

Library of Congress Cataloguing in Publication data
Kwok, S. (Sun)
Cosmic butterflies: the colorful mysteries of planetary nebulae/Sun Kwok.
p. cm.
Includes bibliographical references (p.).
ISBN 0 521 79135 9
1. Planetary nebulae. I. Title.
QB855.5 .K957 2000
523.1'135–dc21 00-068867

ISBN 0 521 79135 9 hardback

Contents

Preface

Planetary nebulae are among the most beautiful objects in the sky. For most of the past century, they have been the favorite observing targets of astronomers, both professional and amateurs alike. Their rich colors and variety of shapes have fascinated generations of sky watchers. Although it has been 200 years since the discovery of the first planetary nebula, their true nature and origins were only ascertained recently. This book is an attempt to provide an accurate and up-to-date account of our understanding of planetary nebulae and how they relate to the death process of stars. In particular, I want to talk about the people and the events that led to our current understanding of this subject. By telling this story, I hope to share some of the excitement and passions that professional planetary nebulae researchers have for this subject with our readers.

One of the interesting properties of planetary nebulae is that they radiate in all parts of the electromagnetic spectrum, from radio to X-ray. Although they have been extensively studied by ground-based optical and radio observations for decades, we now know that these observations do not give a complete picture of planetary nebulae. Our modern understanding of the planetary nebulae phenomenon is to a large degree due to observations in the ultraviolet, infrared, and X-ray made by space-based telescopes. In addition to the historical aspects mentioned above, I tried to emphasize the contributions by these orbiting observatories in this book.

The grandest of all the orbiting observatories is of course the *HST*. The contributions made by the *HST* to planetary nebulae research is obvious from the many beautiful *HST* pictures presented in this book. In spite of the tremendous demands made by astronomers working in all fields of astronomy, the Space Telescope Science Institute has allocated some of its precious telescope time to the observations of planetary nebulae. For this, I (and I hope our readers as well) am profoundly grateful.

Although research on planetary nebulae is carried out with all modern ground-based and space-based telescopes, significant work can still be done with modest size telescopes. Consequently, planetary nebulae are studied by astronomers in many countries in the world and can be considered as one of the most international of all scientific disciplines. One of the greatest joys of working in this area of research is the opportunity to meet and work with colleagues in different parts of the world. Through these interactions, I have come to appreciate a wide diversity of approaches to doing science and the benefit of multi-national collaborations.

The book grew out of a number of popular articles I wrote for the *Sky and Telescope* magazine, and I thank Joshua Roth of Sky Publishing for his interest in my writings on planetary nebulae. I thank Fred Lo and the Institute of Astronomy and Astrophysics, Academia Sinica for their hospitality in providing a pleasant environment for me to write this book. Jeff Stoesz, undergraduate research assistant at the Space Astronomy Laboratory of the University of Calgary, processed most of the *HST* images of planetary nebulae presented. A number of the artworks were done by Yeatland Wong and Valerie Torchinsky. My graduate student Kate Yu-Ling Su participated in the analysis of our *HST* observations, and was responsible for the creation of some of the beautiful images of proto-planetary nebulae and young planetary nebulae. Bruce Balick, Adam Block, You-Hua Chu, Trung Hua, George Jacoby and David Thompson have kindly provided images for use in this book. I also thank my Cambridge University Press editor Adam Black for believing in this project, and for his efforts in bringing the concept to reality. The production of this book was capably handled by the CUP team in the UK. Finally, I thank my wife Emily and my daughter Kelly for their support, and in particular my daughter Roberta for her careful readings and valuable comments on the numerous early drafts.

S.K.
Calgary, Canada
December 2000

1 Planetary nebulae – the last hurrah in the life of a star

Stars do not live forever. The Sun, for example, was born approximately four and a half billion years ago and is expected to have another five billion years left in its life. What is the eventual fate of the Sun and stars like it? In popular science fiction, it is commonly believed that stars explode when they die. For example, in the opening scene of the movie *Superman*, Superman has to escape from his home planet because its parent star explodes as a supernova. Such an event, however, is very unlikely. Any star that lives long enough to allow life to develop in its planetary system would probably not become a supernova. The common belief that violent events such as supernova explosions are the norm is a myth. Recent research has shown that the majority of stars (in fact over 95%) will not explode. Our Sun, being a typical star, will first expand in size, its surface reaching the orbit of the Earth. All life on Earth as we know it will cease to exist. However, in the five billion years before this happens, humans will probably have developed the technology to transport the entire Earth's population to the planetary system around a nearby star. If so, we will be witnessing a spectacular show when we look back at Earth. Gas and dust will begin to stream from the surface of this bloated Sun, creating a gaseous entity that is 500 times larger than the orbit of Pluto. This nebula will give a magnificent display of many-colored light that will last over ten thousand years. This firework show, known as a 'planetary nebula', will be the last hurrah of the Sun before it finally

Figure 1.1.
[opposite] M27, the Dumbbell Nebula in Vulpecula is the largest planetary nebula in the sky. It has an angular size of 8 by 6 arcmin. This picture was taken with the 0.4-m telescope at *Kitt Peak National Observatory* with three broad (red, green, and blue) filters and the colors shown should resemble the perception of the human eye (credit: Adam Block/NOAO/AURA/NSF).

fades away into oblivion. Looking back at the solar system from our new home, we can at least take comfort in a glorious end.

Planetary nebulae are among the most beautiful objects in the sky and have fascinated professional and amateur astronomers alike. Planetary nebulae have elegant symmetries and rich colors. However, the origin and nature of planetary nebulae were not understood until very recently. In this book, we will follow the story of planetary nebulae and see how they fit into the eventful lives of stars.

The first recorded observation of a planetary nebula was by Charles Messier on July 12, 1764, and the object was given the number 27 in his catalogue of nebulous objects. M27, now commonly referred to as the Dumbbell Nebula, is the largest and one of the most easily recognizable planetary nebulae in the sky (Fig. 1.1). The second planetary nebula to be discovered was the Ring Nebula in Lyra (Fig. 1.2), observed by Antoine Darquier of

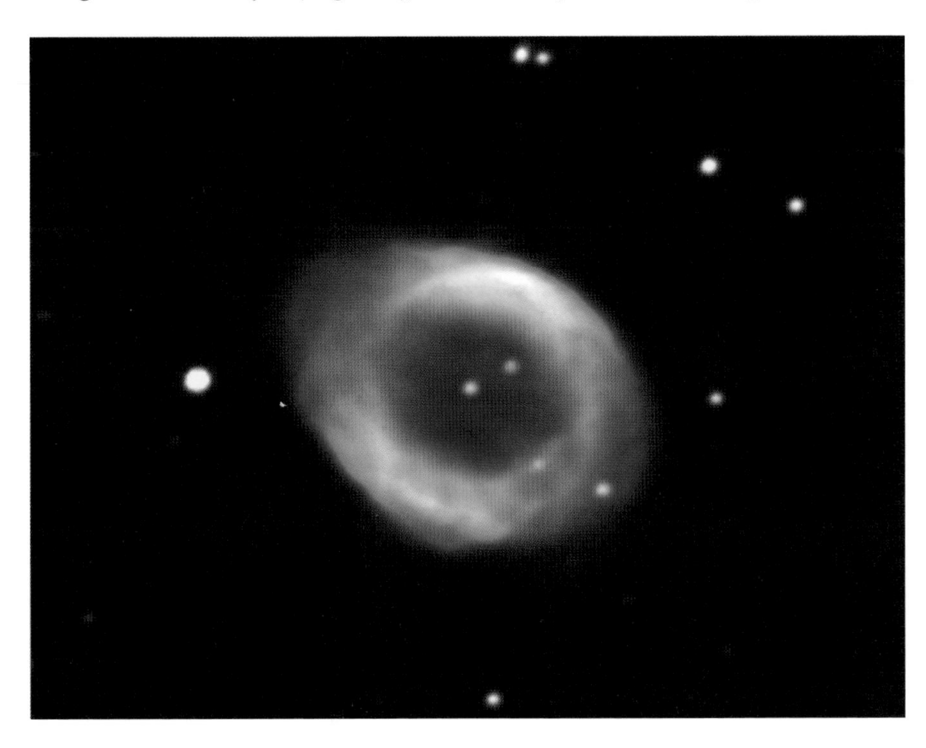

Figure 1.2.
The Ring Nebula in Lyra is probably the most photographed object in the night sky. It has an approximately elliptical shape, with a semi-major axis radius of 42 arcsec and an ellipticity of 1.31. The 15th magnitude star at the center of the ring is the energy source of the nebula (credit: Adam Block/NOAO/AURA/NSF).

Toulouse in January 1779. Darquier described it as 'very dim but perfectly outlined ... it is as large as Jupiter, and resembles a fading planet'. This was followed by the discovery of the Little Dumbbell Nebula M76 in Perseus in September 1780, and the Owl Nebula M97 (Fig. 1.4) in Ursa Major in February 1781 by Pierre Mechain. It is interesting that Messier, a comet hunter, regarded nebulae as a nuisance; the purpose of his catalogue of nebulous objects was intended to avoid misidentifying nebulae as new comets. He might

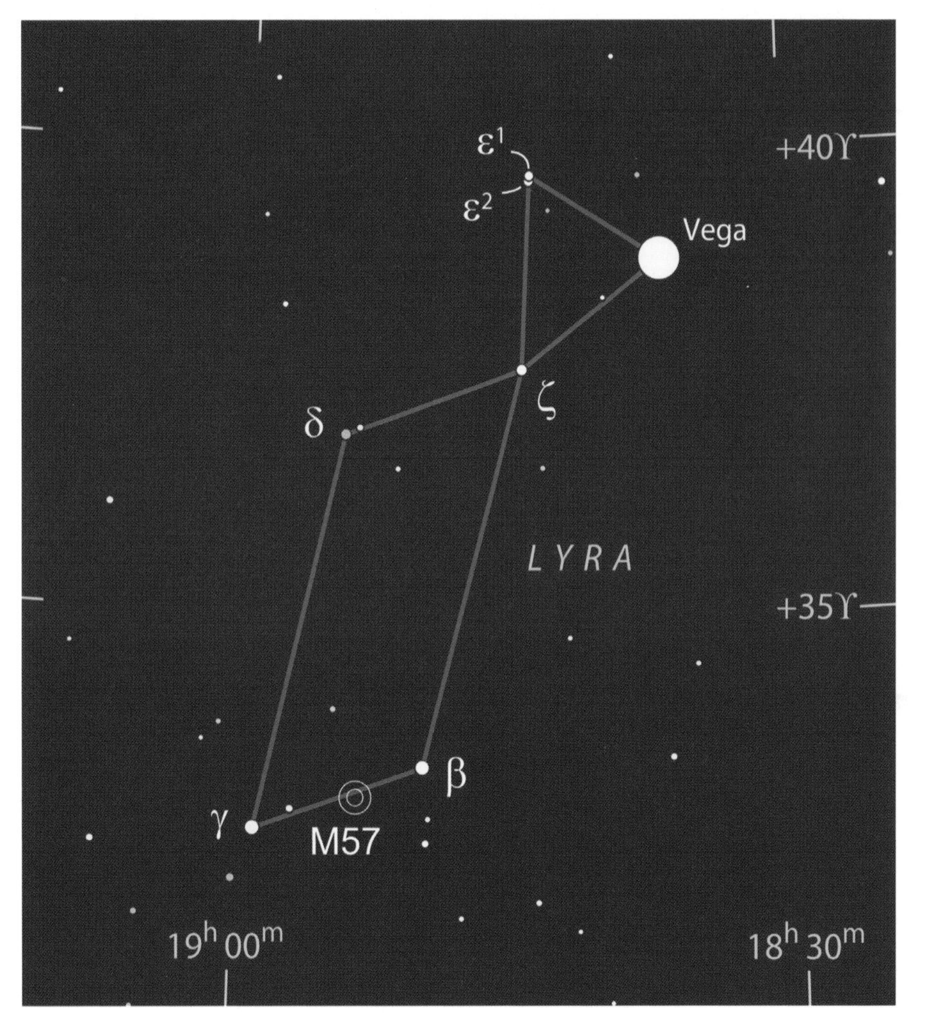

Figure 1.3.
The Ring Nebula, with an integrated visual magnitude of 8.8, is a favorite summer object for northern hemisphere observers. Its ring shape can easily be discerned by small telescopes. This chart is a guide to locating it in the constellation of Lyra.

Figure 1.4.
[opposite] The Owl Nebula (M97) was discovered by Pierre Mechain on February 16, 1781. It was named the 'Owl' by Lord Rosse in 1848 for its pair of 'eyes', although the resemblance is not so strong when viewed by a modern telescope (credit: NOAO).

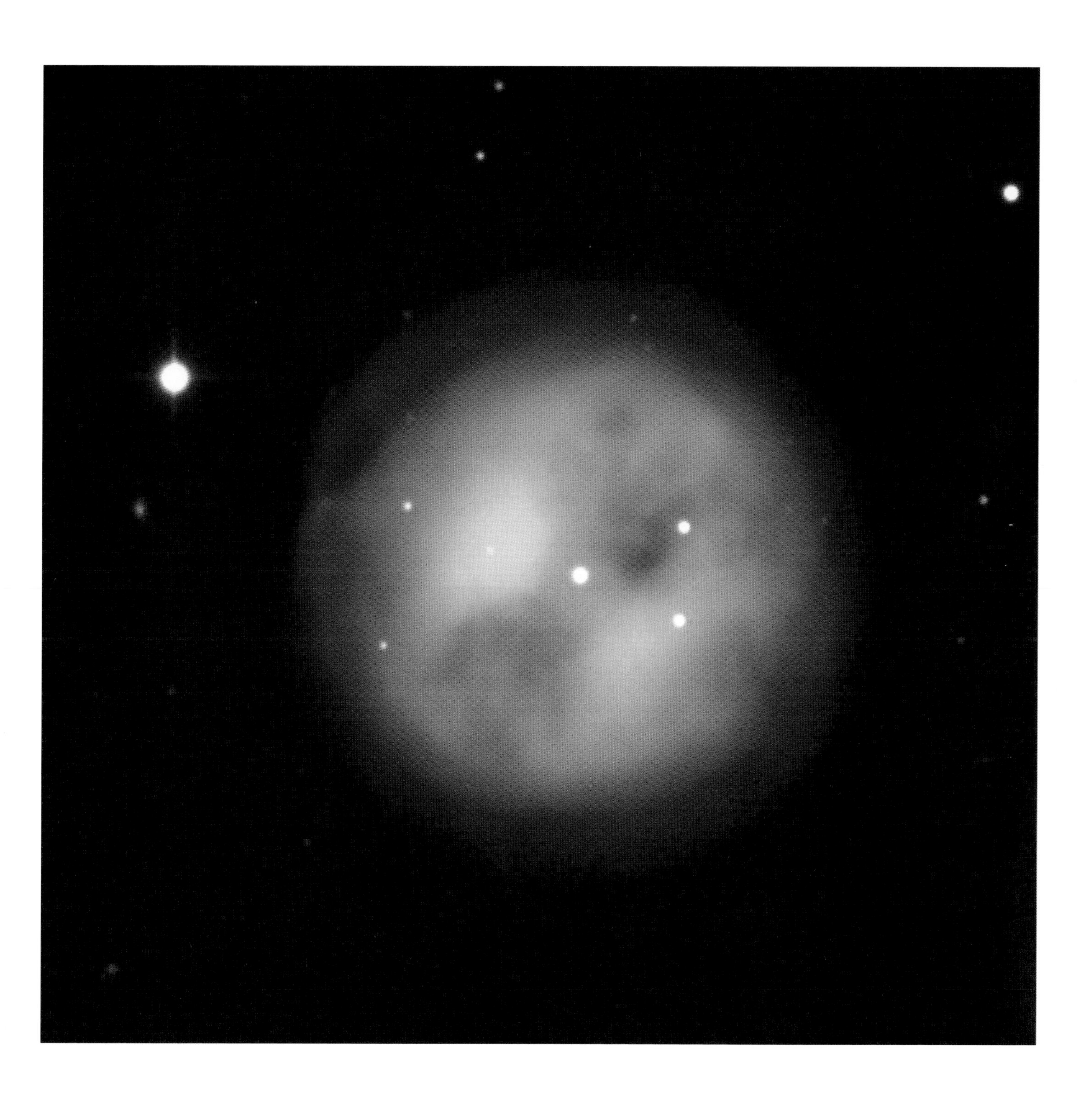

have been amused to find that the nebulous objects in his catalogue (galaxies, star clusters, planetary nebulae, etc.) became the focus of astronomical research in the twentieth century.

The final version of the Messier catalogue of 1784 included four planetary nebulae (M27, M57, M76, and M97). The term 'planetary nebula' was coined by William Herschel (1738–1822), who found that their appearance resembles the greenish disk of planets such as Uranus or Neptune. This property separates them from 'white nebulae', which are made of stars and are now called galaxies. Planetary nebulae are also different from large diffuse nebulae such as the Orion Nebula in that they often have a well-defined symmetrical appearance. This is exemplified by the Ring Nebula

Figure 1.5.
[opposite] NGC 2392 is called the Eskimo Nebula because early low-resolution pictures resemble a human head with a fury hat. This *Hubble Space Telescope* (*HST*) image reveals much more detail of the nebula than previous ground-based images.

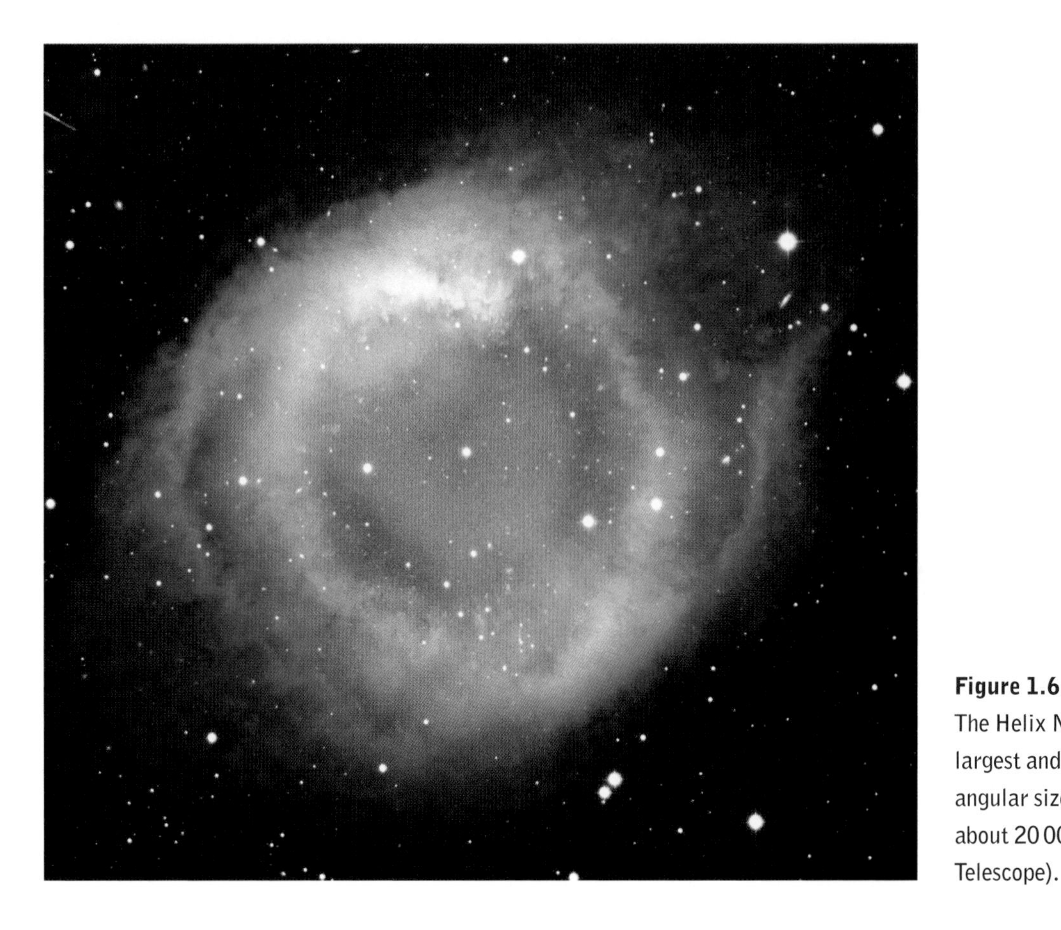

Figure 1.6.
The Helix Nebula (NGC 7293) is among the largest and oldest planetary nebulae. It has an angular size of 6 arcmin and is believed to be about 20 000 years old (credit: Anglo-Australian Telescope).

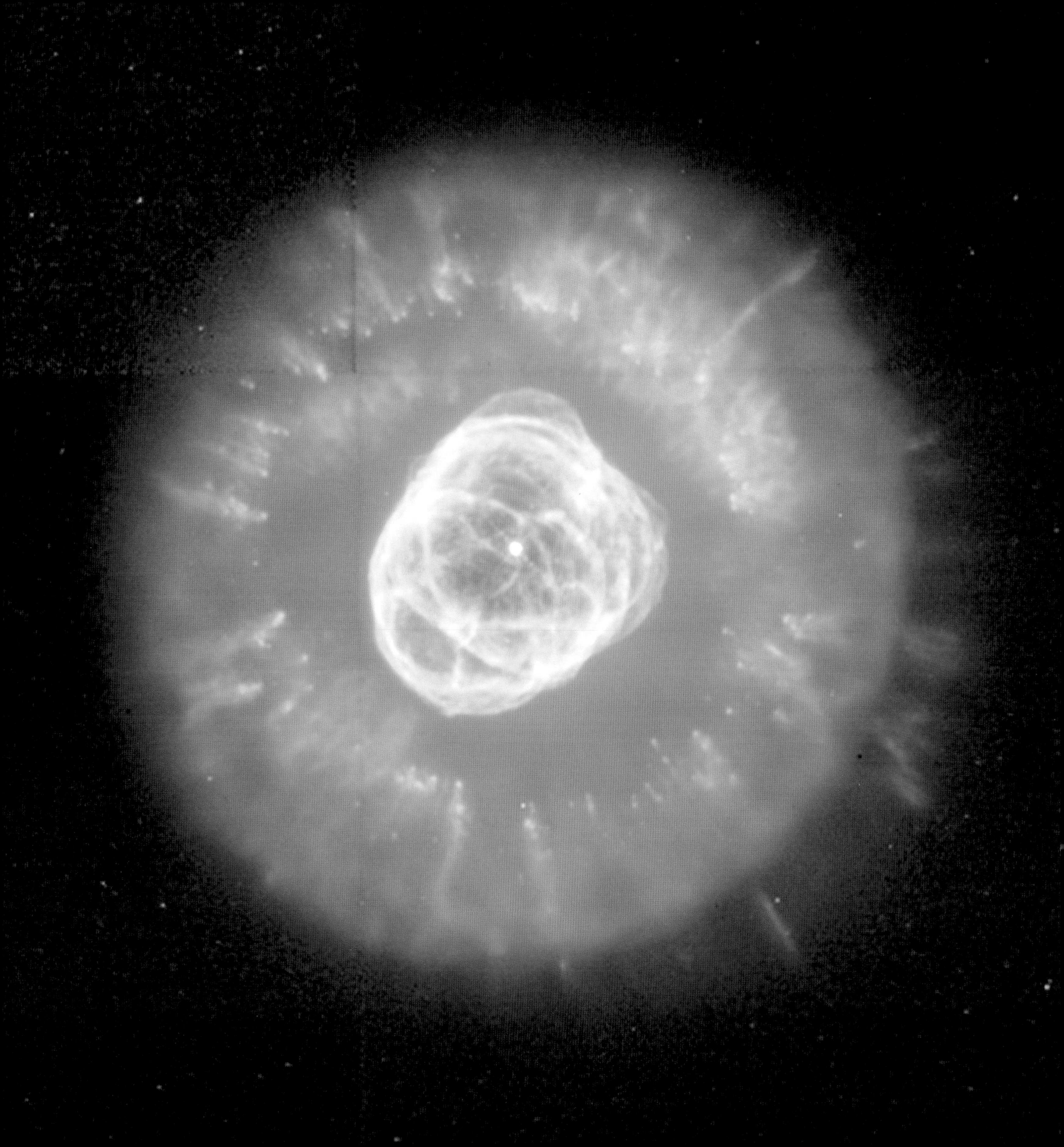

Figure 1.7.
Sir William Herschel (1738–1822) is remembered as the discoverer of the planet Uranus. For this discovery, he received a royal appointment from King George III and a grant of £200 a year. With another grant of £4000, Herschel built what was then (in 1789) the world's largest telescope, a reflector 40 feet in length with a 48-inch mirror. With an earlier 20-foot reflector of 12-inch aperture, he was able to resolve many of Messier's nebulae into stars. He then conjectured that, with larger and larger telescopes, all nebulae could be shown to be made up of stars. The only exception to his theory is planetary nebulae. If the nebulosity is due to a concentration of distant stars, then these stars must be much smaller than the central star. Herschel therefore concluded planetary nebulae are not stellar but are made of a 'shining fluid' of unknown nature.

shown in Fig. 1.2. It is symmetric in the sense that there are at least two lines upon which one can fold the picture and come up with overlapping images.

Many of the brighter planetary nebulae were discovered by William Herschel (Fig. 1.7) and his son John Herschel. *The New General Catalogue of Nebulae and Clusters of Stars* published by Johann Louis Emil Dreyer in 1888 contains many entries of planetary nebulae, some of which are listed in Chapter 21. After his discovery of the planetary nebula NGC 1514, William Herschel

Figure 1.8.
[opposite] The fact that planetary nebulae often have a bright central star convinced Herschel that the nebulous nature of planetary nebulae is different from that of galaxies and star clusters. The planetary nebula NGC 6369 in Ophiuchus contains a bright central star and has an almost perfectly circular shell.

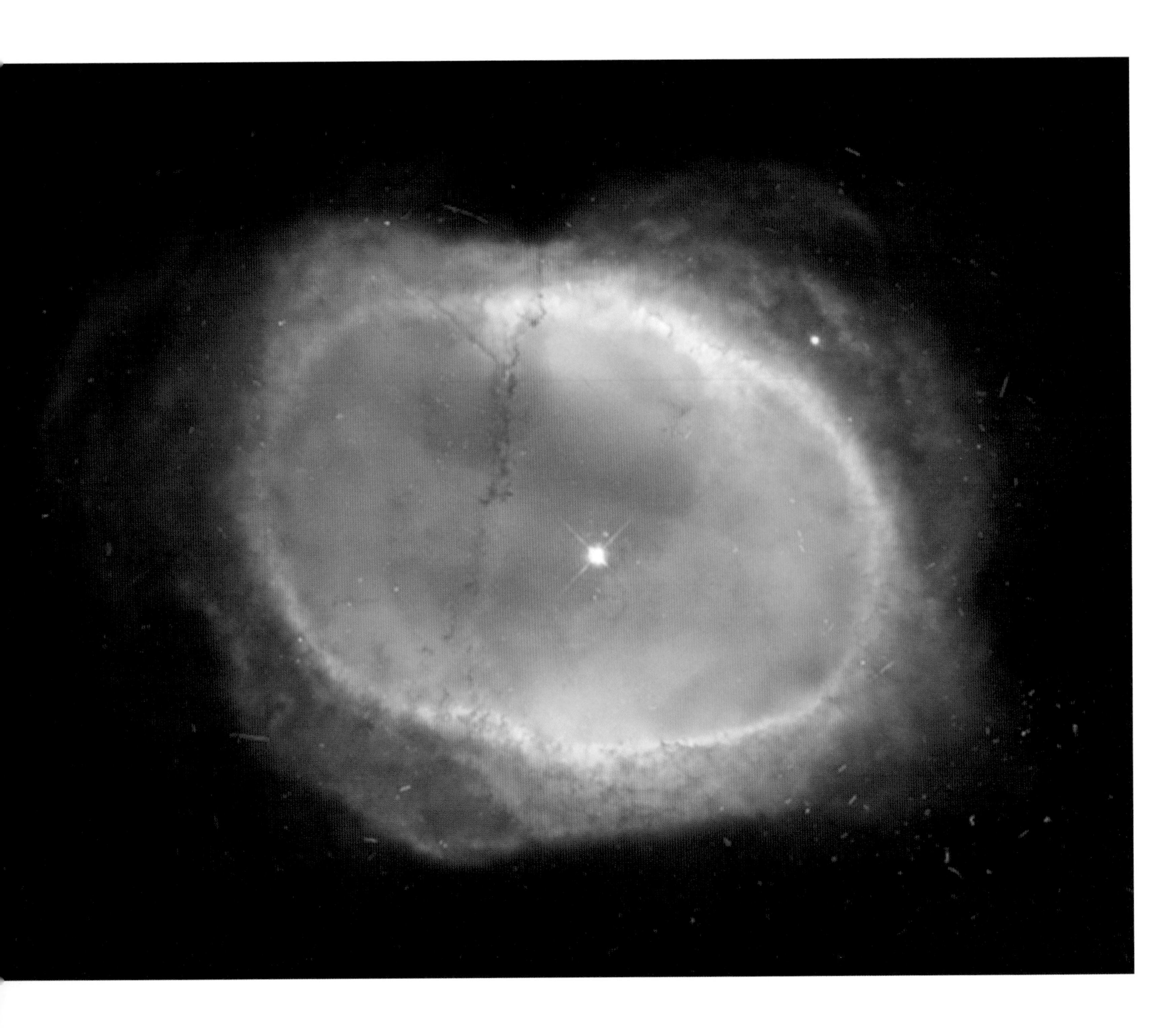

Figure 1.9.
[opposite] The faint star in the middle of the Southern Ring Nebula (NGC 3132), not its bright neighbor, is the star that ejected the nebula. Central stars of planetary nebulae are hot and often faint in visible light.

noted that it has a very bright central star. This convinced him that planetary nebulae are not unresolved stellar clusters, but instead are made up of nebulous (gaseous) material. In 1922, Edwin Hubble found that larger nebulae seem to have brighter central stars, and concluded that the nebulae extract their energy from the star. This one-to-one relationship between the nebula and a central star makes planetary nebulae distinct from other nebulae (see Figs. 1.8–1.10). The nebulous nature of galaxies is due to the superposition of many, many stars, and the large diffuse nebulae like Orion also have many stars associated with them. Planetary nebulae are in a class of their own.

Beginning in the early 1950s, George Abell used the photographs taken by the *National Geographic Society–Palomar Observatory Sky Survey* to discover new large and faint planetary nebulae. However, the technique of visual inspection of the photographic plates only works for objects larger than about 15 arcsec; any object smaller than this limit would be hard to distinguish from ordinary stars. Since the light from planetary nebulae is made up of sharp discrete lines and not continuous white light as in the case with stars (see Chapter 3), they can be identified by placing a prism in front of a telescope. Ordinary stars will show up as a band of continuous colors, whereas planetary nebulae will have sharp bands in the red and in the green. Many planetary nebulae were discovered by this technique: first at *Harvard Observatory* from 1890 to 1910, and later in the early 1960s by Rudolph Minkowski with the 10-inch telescope at *Mount Wilson Observatory*. The same camera was moved to South Africa by Karl Henize where he found more new planetary nebulae in the southern sky. Additional objects were found by the Czech astronomer Lubos Kohoutek and by the Indonesian astronomer Pik Sin The. The first comprehensive catalogue of planetary nebulae was compiled in 1964 by Lubos Perek and Lubos Kohoutek and contained over 1000 entries. This was followed by the *Strasbourg–European Southern Observatory*

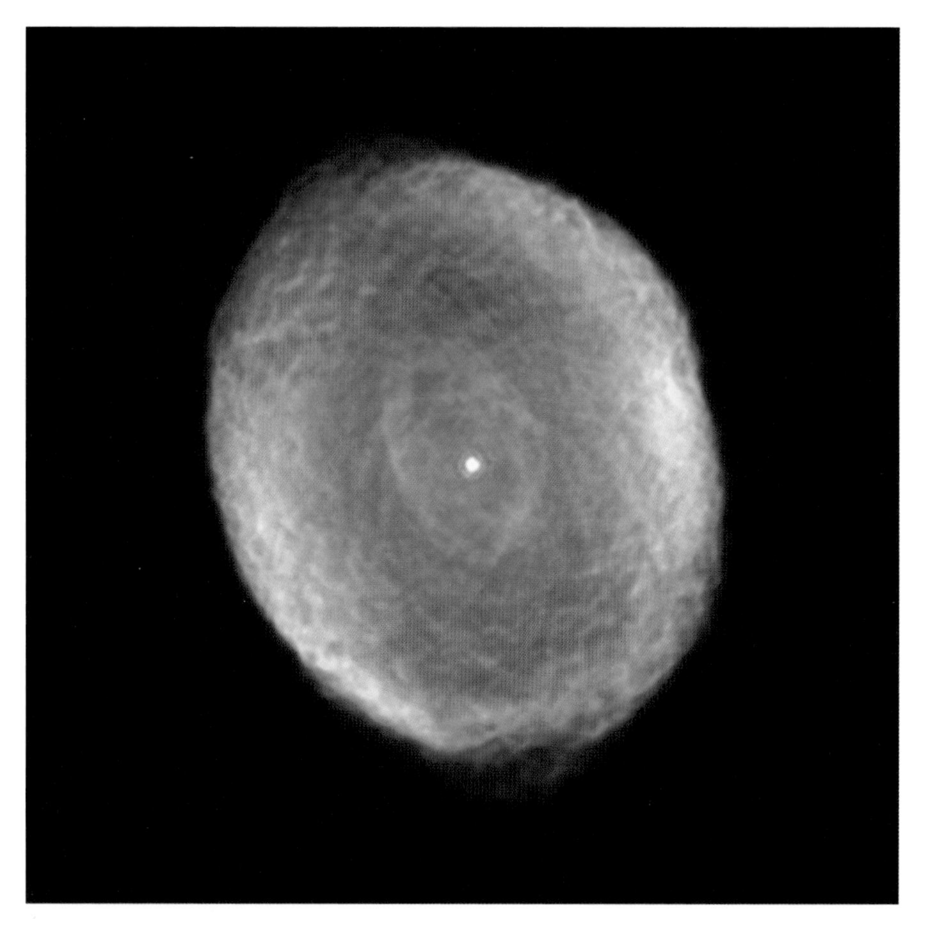

Figure 1.10.
This small (~12 arcsec) planetary nebula (IC 418) in the constellation of Lepus is one of the most extensively studied by astronomers. The bright 10th magnitude central star is the sole source of energy of the nebula (credit: NASA and the Hubble Heritage Team, Space Telescope Science Institute).

Catalogue in 1992 by Agnès Acker and her colleagues. The *Anglo-Australian Observatory/United Kingdom Schmidt Telescope* survey of the southern galactic plane added several hundred more. At the end of the twentieth century, we knew of about 1500 objects in the Milky Way Galaxy that are confirmed planetary nebulae.

2 The shapes and colors of planetary nebulae

While we commonly identify planetary nebulae by their characteristic ring-like shells and symmetrically placed central stars, planetary nebulae actually come in a variety of shapes and forms. Many have elliptical forms, such as NGC 6891 (Fig. 2.1) and IC 2165 (Fig. 2.2), and some have butterfly-like bipolar forms such as NGC 6302 (Fig. 2.3) and M2-9 (Fig. 2.4). Some even have a double-ring structure as seen in the Etched Hourglass Nebula

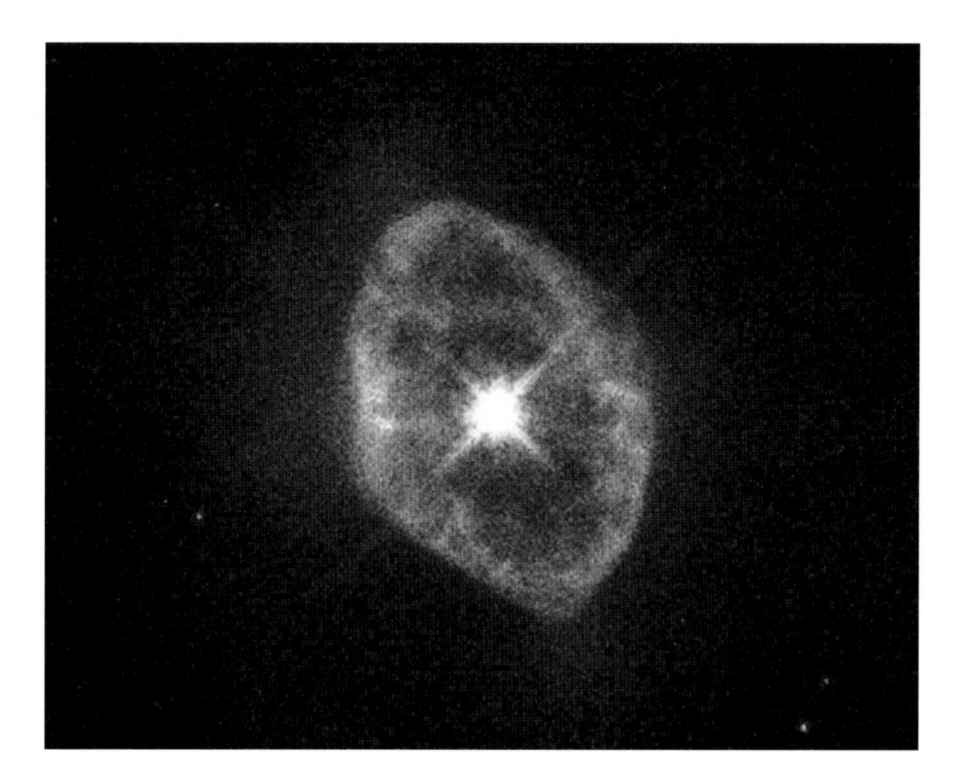

Figure 2.1.
The morphology of planetary nebulae is characterized by a shell of high-density gas surrounding a central star, as in this *HST* image of NGC 6891.

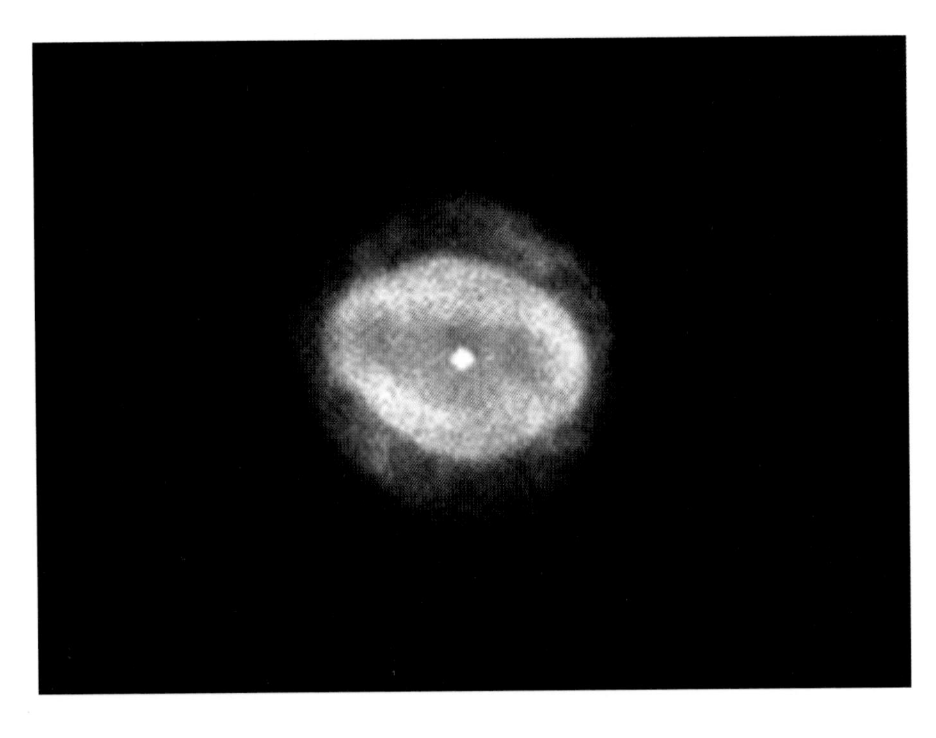

Figure 2.2.
The *New General Catalogue of Nebulae and Clusters of Stars* (NGC) and its supplement *Index Catalogue* (IC) contain many planetary nebulae. The symmetric shape of the nebula and its centrally placed star were essential criteria of the planetary nebula designation of IC 2165.

Figure 2.3.
[opposite] This color image of NGC 6302 is a composite of broad-band blue, yellow, and red images (each from 10-minute exposures) obtained with the 8.2 m *Very Large Telescope of European Southern Observatory* during its commissioning phase in May 1998 (credit: ESO).

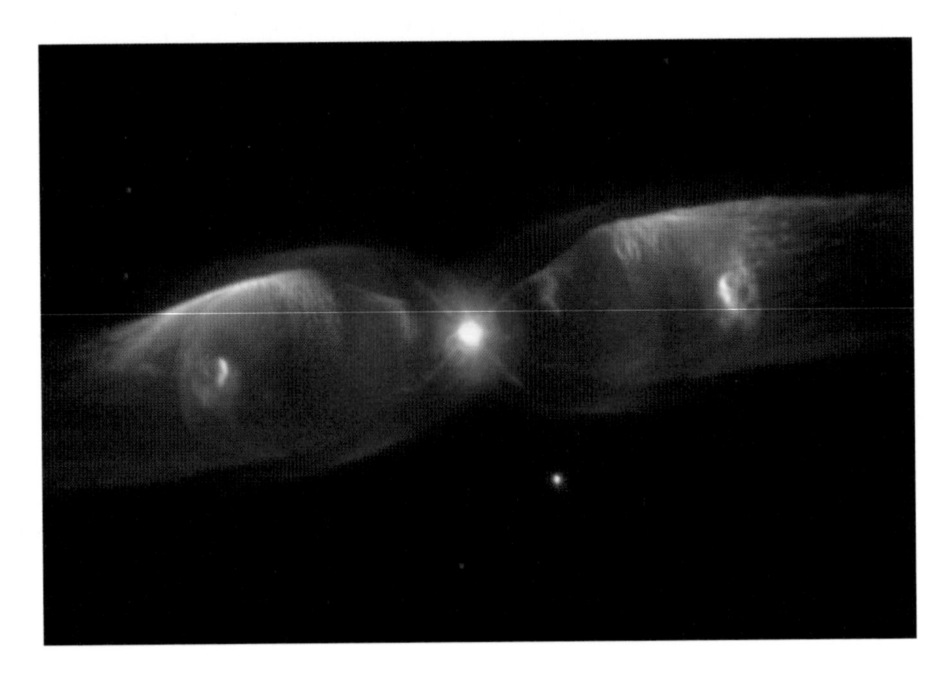

Figure 2.4.
The Butterfly Nebula M2-9 in Ophiuchus was discovered and named by Minkowski. Its cigar-shaped lobes extend over 1 arcmin on each side and are oriented nearly north–south. In this *HST* WFPC2 image, we can see two bipolar lobes, one inside the other, as well as bright knots. One of the most remarkable properties of this nebula is that some of the bright knots have moved since its discovery in 1947, suggesting that they are undergoing a spiral motion within the nebula.

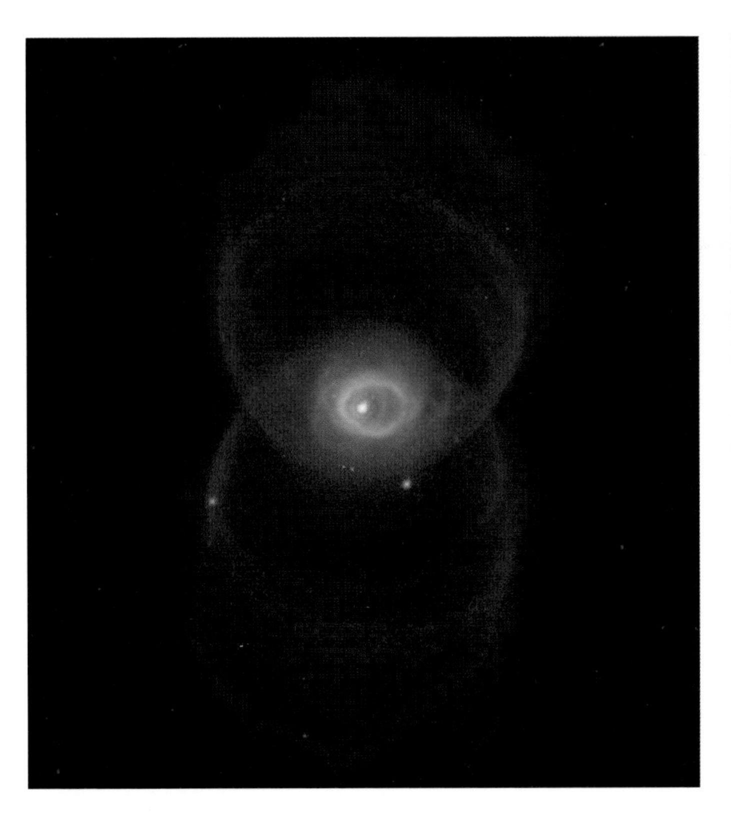

(Fig. 2.5). Planetary nebulae that are perfectly round like Shapley 1 (Fig. 2.6) and Wray 17-31 (Fig. 2.7) are exceptions rather than the rule. In 1918, Heber Doust Curtis of *Lick Observatory* was the first to arrange planetary nebulae in different classes based on their appearances. Since that time, astronomers have used a variety of schemes to classify planetary nebulae morphology, using terms such as stellar, round, elliptical, rings, bipolar, and butterfly.

In contrast to stars, which are primarily white, planetary nebulae have dramatic colors. This is because the nature of light emitted from planetary nebulae is fundamentally different from that of starlight. The fact that sunlight is made up of lights of different colors was demonstrated by Isaac Newton in 1666 using a prism. The spread of colors (known as a spectrum) ranges from violet to red. This is also evident from looking at a rainbow, where

Figure 2.5.
[above left] Dubbed the Etched Hourglass Nebula, MyCn 18 in the constellation of Musca is a striking example of the exotic shapes planetaries can develop. A dense cloud of dust girdling the central star's equator may have thwarted expansion, resulting in a pinched waist.

Figure 2.6.
[above right] Shapley 1 in the southern constellation Norma has a remarkably spherical shape that is unusual among planetary nebulae. A 14th magnitude star sits in the center of the 76 arcsec nebula (credit: Anglo-Australian Observatory, David Malin).

Figure 2.7.
The planetary nebula Wray 17-31 is located in the rich southern constellation of Vela. As part of his PhD thesis research at Northwestern University, James D. Wray identified this object from the southern survey plates obtained by Karl Henize. It has a diameter of about 2 arcmin and a roughly spherical shape. Because of its faintness and southern location, not much is known about this nebula (credit: Anglo-Australian Observatory, David Malin).

water vapor in the atmosphere acts like a prism. A careful examination of the solar spectrum, however, shows that there are sharp breaks among the continuous colors. These dark lines were found to be due to absorption by atoms in the solar atmosphere. The helium atom, the second lightest element after hydrogen, was the first element discovered not in a terrestrial laboratory but in the spectrum of the Sun. Using more powerful instruments called spectrographs, astronomers have now tabulated thousands of lines due to different elements in the solar spectrum.

The first spectrum of planetary nebulae was taken by Sir William Huggins on August 29, 1864. The target of his observation was

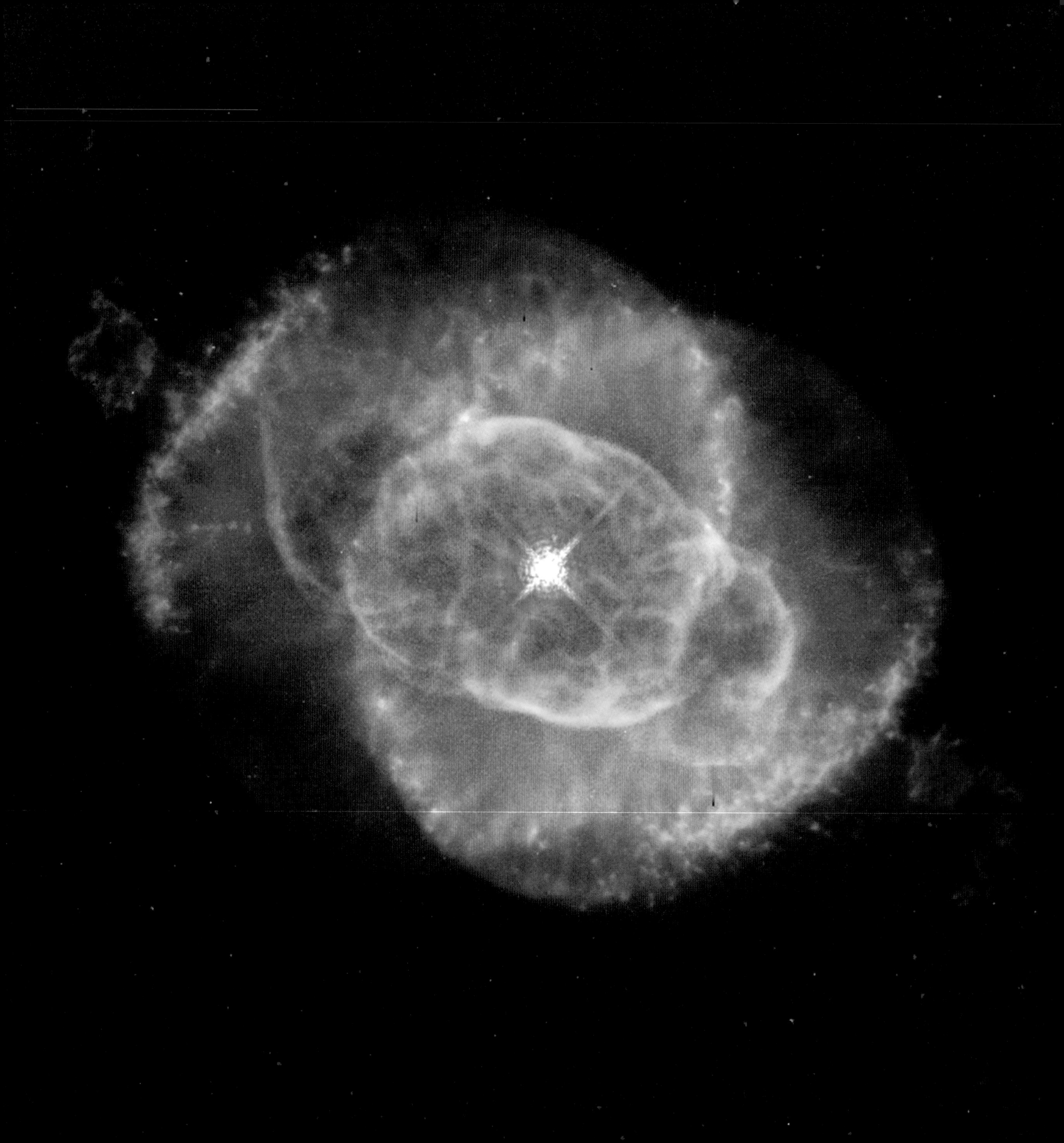

Figure 2.8.
[opposite] NGC 6543 in Draco was the first planetary nebula to have a spectrum taken. The *HST* WFPC2 image shows a crossed pair of ellipses in the center of the nebula, therefore giving the name 'Cat's Eye Nebula'.

NGC 6543 in the constellation of Draco (Fig. 2.8). Instead of the continuous light seen in the spectra of stars, he was surprised to see just three bright lines: one blue and two green (Fig. 2.9). The blue line was identified as due to the hydrogen atom, but the two green lines had never been seen before, neither in the spectra of stars nor in the laboratory.

These two mysterious green lines caused a stir in the physics community. Astronomers thought they were due to a previously unknown element. Following the precedent of helium, they named this new element 'nebulium'. In 1926, Ira S. Bowen of California Institute of Technology identified the green lines to be due not to some exotic new element but to the common element oxygen. The reason that these lines of oxygen are not seen in the terrestrial laboratory environment is that the density of the Earth's atmosphere is too high. Only under the very low-density conditions of planetary nebulae (see Chapter 3) can these lines arise.

Bowen's discovery kindled a great deal of interest in the field of planetary nebulae. For the next several decades, astronomers worked closely with atomic physicists to analyze the spectra of planetary nebulae in search of other atomic lines unseen on Earth.

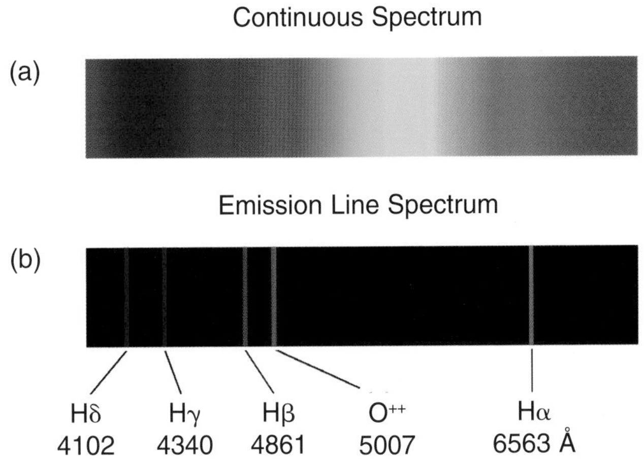

Figure 2.9.
A comparison between a continuous spectrum (upper panel) and an emission line spectrum. The red and blue lines are due to hydrogen. The two mysterious green lines at 4363 and 5007 Å first labeled nebulium were later identified as due to ionized oxygen.

The quantum theory, which was developed in the 1920s, was applied to understand the nebular spectra. By combining the theoretical knowledge of quantum physics with the observed brightness of the nebular atomic lines, astronomers can determine the temperature and density of the nebulae as well as the quantity of each atomic element relative to hydrogen. Donald H. Menzel of *Harvard Observatory* was a pioneer in the field. His work was developed by his student Lawrence Aller, who later became a professor at the University of Michigan and UCLA. Aller, who did extensive spectroscopic observations and theoretical interpretations of such spectra, is responsible perhaps more than anyone for the development of modern physics of gaseous nebulae. The application of quantum physics to planetary nebulae is in part responsible for changing astronomy from a purely observational science to the modern discipline of astrophysics.

3 How do planetary nebulae shine?

Let us pretend for the time being that we choose to build a human settlement in an artificial space station located about a light year away from the Sun. We would avoid being engulfed by the swelling Sun, but would be within reach of the expanding planetary nebula. What would it be like inside a planetary nebula? Astronomers, using the brightness of atomic lines as thermometers, can accurately measure the temperature. We find that the typical temperature inside a planetary nebula is about 10 000 °C (degrees Celsius). Although this is much higher than the comfort limit of about 30 °C that we are accustomed to on Earth, the good news is that we wouldn't feel it because of the low density. The density of a planetary nebula is about 100 000 000 000 000 000 (yes, 17 zeros) times lower than the density of the Earth's atmosphere at sea level. This is much better than any vacuum that scientists could create in an Earth-based laboratory. Inside such a vacuum state, the temperature of the 'air' is irrelevant because it is not the sole source of heat. As people who live in cold countries like Canada or Russia can testify, on some winter days when the air temperature is −30 °C, one can in fact feel 'warm' because of the bright sunshine. This is the case inside a planetary nebula: the radiation from the star in the center of the nebula has more to do with how we would feel than the 'air'. And the star is HOT. Typical temperatures of the central stars of planetary nebulae range from 30 000 to 150 000 °C, or 5–25 times hotter than the Sun. In fact, central stars of planetary nebulae are the hottest stars in the Universe (see Fig. 3.1).

Figure 3.1.
[opposite] Not all central stars of planetary nebulae are easy to see. Stars that are very hot radiate primarily in the ultraviolet and can be very faint in the visible. NGC 2440, whose central part is shown in this *HST* image, has one of the hottest central stars with a temperature of ~200 000 °C – more than 30 times hotter than that of our own Sun.

How much heat would you feel from such hot stars? This depends on how far away you are. We know from experience that the farther you sit from a fireplace or campfire, the less you feel the heat. In the solar system, the Earth is about 8 light-minutes away from the Sun and the temperature is just 'right' (this is, of course, why we are here). But Jupiter, which is 43 light-minutes away, has a temperature of about −150 °C. The usual separation between a planetary nebula and its central star is about 1 light-year. So, in spite of the hot central star, you wouldn't feel that hot. The temperature that you would feel from the radiation of the central star is about −150 °C. This may be cold by Earth standards, but is in fact quite warm in comparison to the temperatures found in dark interstellar space, about −250 °C.

The danger that one faces from the hot central star is not the *amount* of radiation but the *kind* of radiation. The Sun, having a temperature of about 6000 °C, emits most of its light in the visible. Visible light, by definition, is the kind of light that human eyes can see. (This is not a coincidence. Over millions of years of evolution, most animals develop vision that makes the best use of the available light. If we had X-radiation vision, it would not be very useful because the Sun does not emit much X-ray and we would be in the dark all the time!) Central stars of planetary nebulae, which are hotter than the Sun, radiate a lot of ultraviolet light. Ultraviolet light carries more energy than visible light and is dangerous to your health. It has the power to break up the chemical bonds of molecules and therefore can destroy living cells. In fact, there is so much ultraviolet radiation coming from the central star that no molecules are expected to survive in planetary nebulae (however, see Chapter 16).

The ultraviolet light not only breaks up molecules but also strips off electrons from atoms. For example, the oxygen atom, which normally has eight electrons, can lose up to two electrons due to ultraviolet light. Hydrogen atoms, the most common atom

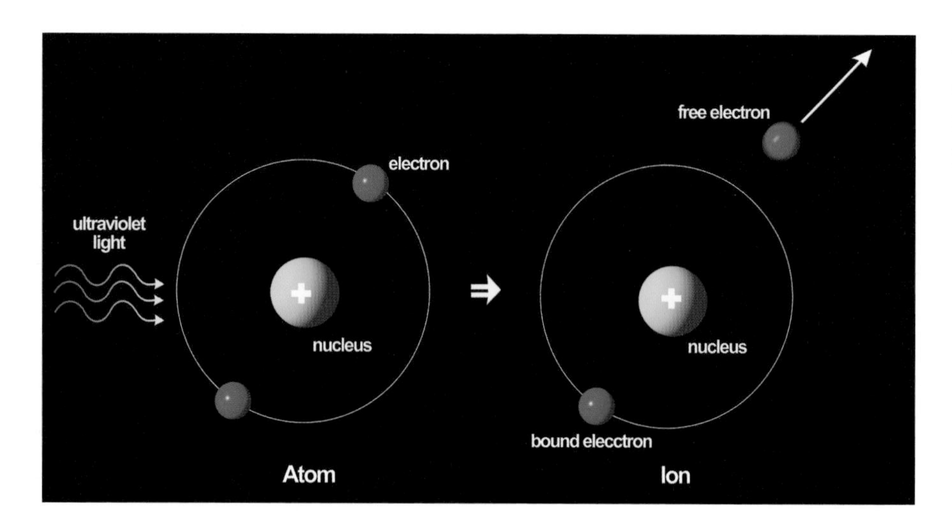

Figure 3.2.
A schematic diagram showing an electron being removed from the orbit of an atom as the result of the atom being hit by an ultraviolet photon.

in planetary nebulae, will lose their only electron, leaving behind a bare proton. Atoms that lose one or more electrons are positively charged, and physicists call this an ionized state of matter (Fig. 3.2).

From time to time, the negatively charged electrons will be attracted to the positively charged protons, reuniting as a hydrogen atom. In the process of reunification, a series of light with specific wavelengths is emitted. Sometimes the free electrons can collide with an atom (e.g., oxygen, nitrogen, sulphur, etc.) and cause an electron inside the atom to be excited to a higher orbit. When the electron falls back down to a lower orbit, light of a specific wavelength is again emitted. Astronomers call this 'line radiation', in contrast to the continuous light emitted by stars. The difference between line and continuous radiation is analogous to the difference between fluorescent lights and incandescent light bulbs. Light bulbs emit continuous light of many colors (together making it white), whereas fluorescent tubes and neon lights emit radiation of discrete wavelengths. So planetary nebulae are just giant neon signs in the sky. These neon lights draw their power from the central star. But instead of having wires to connect the nebula and the central star, the star uses its ultraviolet light to 'light up' the nebula.

Planetary nebulae not only draw attention to themselves with their neon signs, they are also broadcasting loudly in the radio. When the free electrons in the nebula pass by a proton, radio waves are emitted. After the Second World War, scientists who worked on radar during wartime used their equipment to study the Universe. With radio receivers, astronomers were able to detect radio signals from a variety of celestial objects, including planetary nebulae. Among the early pioneers was Yervant Terzian of Cornell University. Beginning in the 1960s, Terzian used the radio telescopes at the *National Radio Astronomy Observatory* in Green Bank, West Virginia, to study planetary nebulae. Complementary work in the southern hemisphere was done by Doug Milne of the CSIRO in Australia, who used the 64-m telescope at Parkes, New South Wales, to measure the radio signals from planetary nebulae. Working with Lawrence Aller, Milne compiled an extensive catalogue of radio measurements from which the density and mass of the nebulae can be derived.

Because radio waves have much longer wavelengths than visible light, the quality of the images one obtains is much worse than that in the optical. One way to compensate for this is to build larger radio telescopes. However, because the steel structures are heavy, there is a practical limit on how large a telescope one can build. Currently, the largest steerable radio telescopes are about 100 m in diameter. Even at that size, the image quality is still poor. In the 1960s, astronomers thought of a way to simulate a large radio telescope by connecting several physically separated telescopes by electric cables. By combining the signals from the telescopes together with a computer, they could construct higher quality images. This technique, known as interferometry, is similar to the computer tomography scans used in medical imaging. The first array of telescopes covering a distance of 5 km was built at the University of Cambridge. In the US, a three-element array was built by the *National Radio Astronomy Observatory* in Green Bank.

Figure 3.3.
The 46-m radio telescope of *Algonquin Radio Observatory* was used by the author to observe planetary nebulae in the late 1970s and early 1980s. It was built deep in Algonquin Provincial Park, Ontario, in order to avoid manmade radio interference. This telescope was closed in 1985 as the result of lack of funding.

Since radio waves generated by planetary nebulae are not affected by dust, which sometimes obscures the visible light, radio observations can provide useful complementary information on the structures of planetary nebulae.

These observations have shown that planetary nebulae are powerful radio sources. NGC 7027, although small and unspectacular in appearance, is one of the brightest radio sources in the sky. From the amount of radio signals it emits, astronomers determined that the nebula of NGC 7027 is one of the most dense of all planetary nebulae. Its high density suggests that it is very young, only about a thousand years old. The ability to distinguish young and old nebulae is important because it allows us to trace their evolution.

4 The young and the old

One of the properties of planetary nebulae is that they are constantly expanding and dispersing into space. According to the principle of the Doppler effect, the wavelength of emission lines will shift to the red or the blue if the atoms that emit the line are moving away from or towards us. Using spectroscopy, astronomers have found that the front and the back sides of the nebular shell are separating from each other at about 40 km/s (or c. 90 000 miles/h). In other words, the shell is expanding from the center at a rate of about 20 km/s. As the nebula expands, its density drops. Consequently, the density in the nebula will eventually be so low that the light emitted from it will be too faint to be seen. The matter in the nebula then gradually mixes with the gas and dust in the general interstellar medium and disappears from view.

If planetary nebulae are expanding at a constant rate, then one can calculate the age of the nebula by dividing its size by the expansion rate. The largest planetary nebula has a radius of approximately 2 light years. Dividing this size by the typical expansion velocity of 20 km/s, we obtain an estimate of the lifetime of planetary nebulae of 30 000 years. While 30 000 years may seem a long time in human terms, this is a blink of an eye when compared with the total stellar lifetime of billions of years.

Even though the lifetimes of planetary nebulae are short, we can still tell the young planetary nebulae from the old ones. Because they are expanding, the young ones are small and bright, while the old ones are large and faint. By searching the Palomar Sky Survey

Figure 4.1.
[opposite] The age of Abell 30 in Cancer is estimated to be 40 000 years, making it one of the oldest planetary nebulae known. This image of A30 was taken by Trung Hua with the 1.2-m telescope of the *Haute Provence Observatory*.

plates, George Abell of *Mount Wilson Observatory* and later of UCLA was able to find many old planetary nebulae. Two examples of old nebulae found by Abell are shown in Figs. 4.1 and 4.2. A planetary nebula older than these would be so faint that it would be beyond the sensitivity limit of current telescopes.

While very old planetary nebulae are hard to find because they are too faint, very young planetary nebulae are also hard to find because they are too small. The fact that turbulence motions in the Earth's atmosphere make images of stars fuzzy does not help. When light from celestial objects travels through the atmosphere, it is distorted so that any object, no matter how small, will seem to have

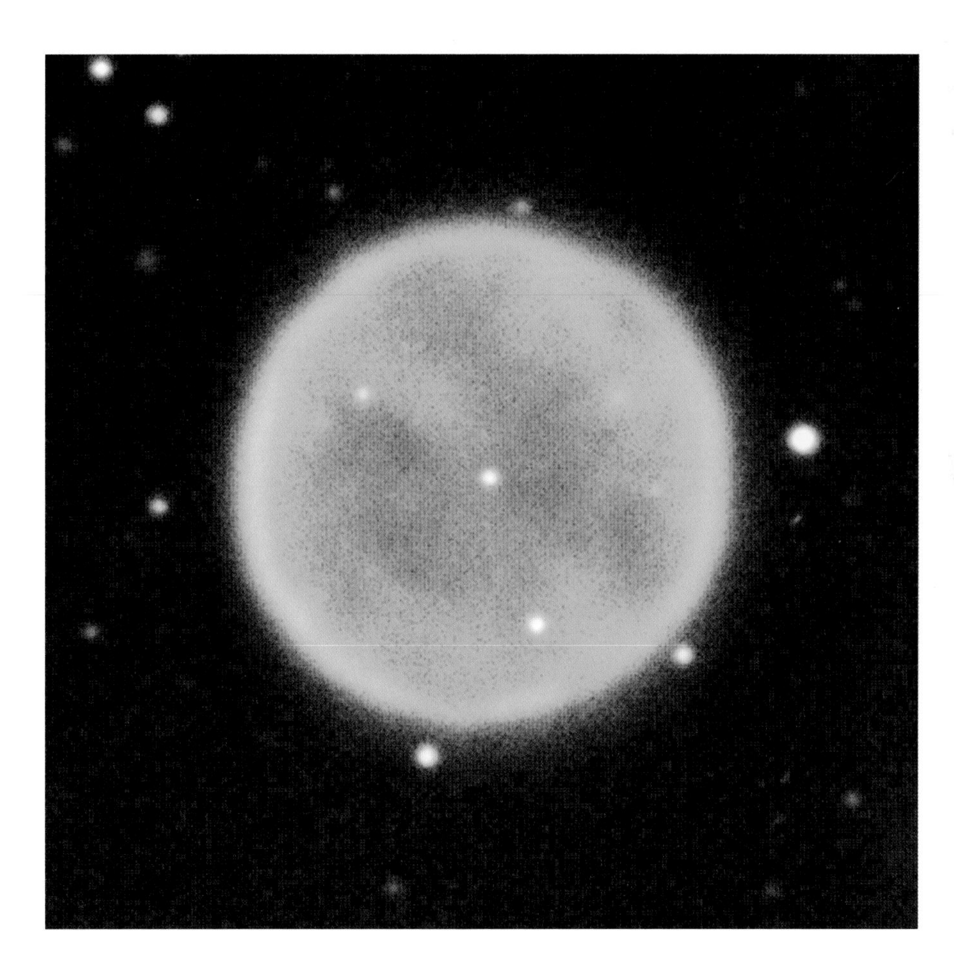

Figure 4.2.
Abell 39 in Hercules has a nearly perfect spherical shape. Like Abell 30, it is one of the old planetary nebulae discovered by George Abell. Photograph by Trung Hua.

Figure 4.3.
The *Very Large Array* is an array of 27 radio telescopes built on the Plain of Augustine (2124 m or approximately 7000 ft above sea level) in New Mexico. Its 27 25-m telescopes are arranged in a Y-shape configuration, giving it an equivalent angular resolution of a 36-km telescope and sensitivity of a 130-m telescope. It was completed in 1980 at a cost of US $78 million. To this day, it remains the world's most powerful radio telescope (credit: NRAO).
In the 1980s, two major radio surveys of planetary nebulae were carried out at the *VLA*. One was by Albert Zijlstra and Stuart Pottasch of the University of Groningen, and the other by Orla Aaquist and Sun Kwok of the University of Calgary. The Groningen group concentrated on larger and more evolved nebulae, and the Calgary group concentrated on more compact and therefore younger nebulae. Together these two surveys provided the most accurate radio images of planetary nebulae.

a minimum angular size of about 1 arcsec. This limit on angular resolution imposed on ground-based optical telescopes makes it impossible to distinguish the shell structure of very young planetary nebulae.

The first hope of finding very young planetary nebulae began in 1980 when the *Very Large Array* (*VLA*, Fig. 4.3) was constructed by the *National Radio Astronomy Observatory* in New Mexico. Using the principle of interferometry, the *VLA* can achieve angular resolution more than ten times better than that of ground-based optical telescopes. Because planetary nebulae are strong radio sources, it is possible to image them with the *VLA*. Since the radio surface brightness is expected to decline as the nebula expands, very young planetary nebulae can be identified by their very high radio surface brightness: in other words, they are radio bright and small. Over a period of several years in the 1980s, my graduate

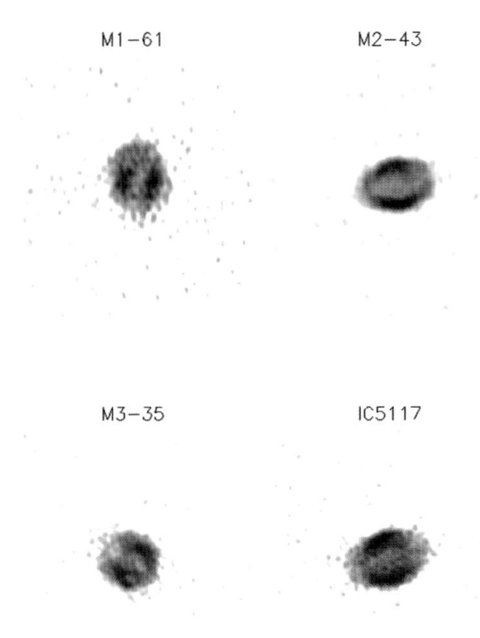

Figure 4.4.
Radio images of four young planetary nebulae (M1-61, M2-43, M3-35, IC 5117) obtained with the *VLA*. The angular resolution in each of these images is about 0.1 arcsec, or about 10 times better than what is possible with ground-based optical telescopes.

student Orla Aaquist and I analyzed images of hundreds of planetary nebulae observed with the *VLA* and were able to identify a group of very young planetary nebulae. Four examples of these radio images are shown in Fig. 4.4.

The problems faced by ground-based optical telescopes can be overcome by placing a telescope in space. The *Hubble Space Telescope* (*HST*) is the dream of optical astronomers come true (Fig. 4.5). Placed at an altitude of 600 km by the space shuttle, the angular resolution of *HST* is only limited by the size of its aperture. With a mirror of 2.4 m, the *HST* can achieve an angular resolution of 0.05 arcsec in the optical. However, when the *HST* was finally launched on April 25, 1990, after a long delay caused by the *Challenger* disaster, NASA found that it suffered from problems with its primary mirror, which was 2 μm (also called micrometer or micron) too flat at the edge. This created an aberration resulting in poor image quality. The disappointment of the astronomical community was indescribable.

Fortunately, the *HST* was designed to be in low Earth orbit so as to allow access by the space shuttle for the repair and replacement of its equipment. NASA followed through with its determination to make *HST* work. On December 2, 1993, astronauts on board the space shuttle *Endeavor* (Figs. 4.6 and 4.7) replaced the original camera by Wide Field Planetary Camera 2 (WFPC2), a spare instrument developed in 1985 by the Jet Propulsion Laboratory in Pasadena, California. A corrective lens was installed in WFPC2 to correct for the defective mirror. With its new 'glasses', the *HST* was finally able to 'see' clearly.

One of the pictures taken by the *HST* during the test of the newly installed WFPC2 was the planetary nebula NGC 2346. This picture of unprecedented clarity fulfilled all the original expectations of the *HST* and greatly excited the planetary nebulae community.

Young planetary nebulae are logical targets of the repaired *HST*. Their relative high brightness requires only very short exposure

Figure 4.5.
[opposite] *The Hubble Space Telescope* in space. The *HST* is in a 600-km-altitude orbit circling the Earth about once every 100 min at a speed of 27 200 km/h (credit: NASA).

Figure 4.6.
[above] The *HST* is held by the Canadarm of the space shuttle *Endeavor* during the first service mission in 1993 (credit: NASA).

Figure 4.7.
[right] *HST* sitting above the space shuttle *Endeavor*'s payload bay against a backdrop of the west coast of Australia (credit: NASA).

times. Using the *HST* in the 'snapshot' mode, Reghvendra Sahai of JPL obtained images of many young planetary nebulae. In Fig. 4.8, we show some examples of *HST* images of young planetary nebulae. Unprecedented degrees of detail can be seen in these images. How planetary nebulae acquire their unique shapes is a fascinating topic which we will address later in this book.

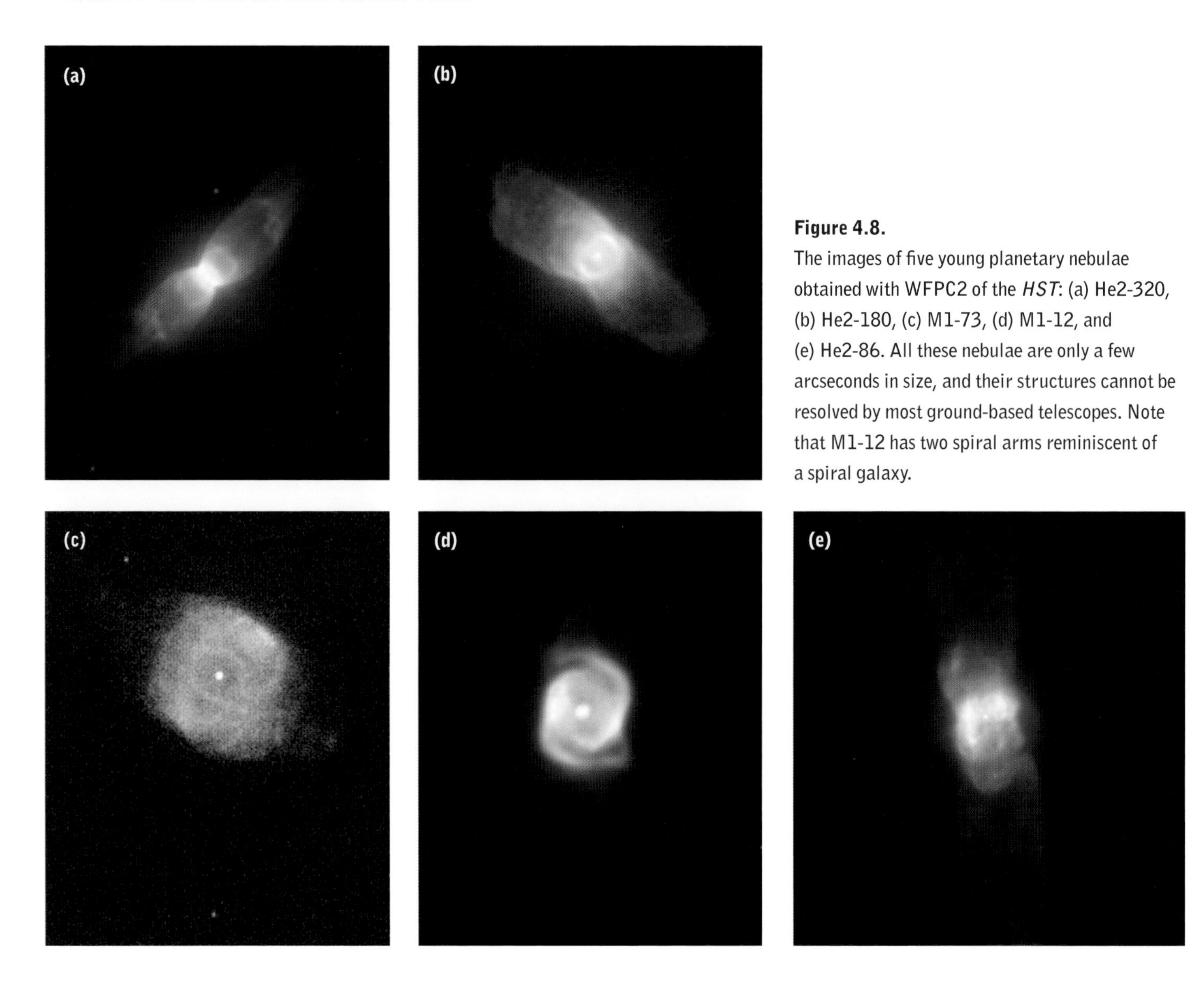

Figure 4.8.
The images of five young planetary nebulae obtained with WFPC2 of the *HST*: (a) He2-320, (b) He2-180, (c) M1-73, (d) M1-12, and (e) He2-86. All these nebulae are only a few arcseconds in size, and their structures cannot be resolved by most ground-based telescopes. Note that M1-12 has two spiral arms reminiscent of a spiral galaxy.

5 Where do planetary nebulae come from and what will they become?

At the beginning of the twentieth century, planetary nebulae were believed to be young stars newly formed in the Galaxy. The conventional wisdom at the time (now recognized to be misguided) was that stars become cooler as they evolve. The central stars of planetary nebulae, being very hot, were therefore thought to be very young stars. One way to test this hypothesis was to find out where planetary nebulae are located in the Galaxy. Astronomers believe that the Galaxy began as a spherical cloud of gas and over time collapsed into its present disc shape. Young stars which were recently born in the Milky Way Galaxy are therefore all concentrated on the Galactic plane (and in particular in the spiral arms), whereas old stars, some born billions of years ago, have larger vertical distances from the plane. Heber Doust Curtis, an astronomer at the *Lick Observatory* in California, studied the distribution of planetary nebulae in the Galaxy and the speed at which they move in the Galaxy, and came to the conclusion in 1918 that planetary nebulae have more in common with old stars than young stars.

The origin of planetary nebulae was a major mystery. If they were evolved stars, where did they fit into the scheme of stellar evolution? In 1939, Hans Bethe of Cornell University identified nuclear fusion as the source of energy of stars. Fusion, the principle behind the hydrogen bomb, works by burning four hydrogen nuclei into one helium nucleus, releasing a large amount of energy. Since stars are made up almost entirely of hydrogen, the amount of fuel available is sufficient for stars to shine for billions of years. The

Sun, for example, is expected to have enough hydrogen fuel for a total lifetime of 10 billion years.

What happens when the hydrogen fuel runs out? Using stars in stellar clusters as a guide, astronomers believe that the Sun will become a red giant. Red giants have lower surface temperatures but are more luminous. They are also larger in size when this transformation takes place. The Sun will increase its size gradually until it is about one hundred times its current size. When viewed from the Earth, the Sun will fill a much larger part of the sky and will take several hours to rise and set.

In 1956, the Russian astronomer Josif S. Shklovsky suggested that a star as large as a red giant would not be very stable. If somehow the atmosphere of a red giant could be detached from its core, then it could flow away from the star and be seen as the shell of a planetary nebula. This view was supported by George Abell and Peter Goldreich in 1966, who found that the expansion velocities of planetary nebulae are consistent with their being thrown out from red giants. If this is true, then what is the force that causes the ejection? We will come back to this question in Chapter 10.

In 1844, F.W. Bessell noted that the brightest star in the sky, Sirius, appeared to be wobbling as if it was under the influence of the gravity of a companion star. However, the companion of Sirius is so faint that it was not discovered until 1862, when Alvan Clark identified it as a faint blue star. Sirius B, as it came to be called, was the first of a newly discovered class of star called white dwarfs.

White dwarfs are different from the normal stars known at that time. Normal, middle-age stars like our Sun are redder in color if they are intrinsically faint. These faint, blue white dwarfs are stars at the end of their lives after they have exhausted all their nuclear fuels. Without nuclear energy, they maintain their luminosity by contraction, gradually using up their gravitational energy reserve. In the process they become fainter and fainter until they can no longer be seen.

Figure 5.1.
[opposite] At a distance of 7000 light years, M4 in the constellation Scorpius is the nearest globular cluster to Earth. Because of its old age (estimated to be 14 billion years), this cluster contains mostly red giants and low mass stars (yellow solar-like stars and red dwarfs) that are yet to become red giants. This picture was taken with the 0.9-m telescope of *Kitt Peak National Observatory* (credit: M. Bolte).
A small portion of the cluster was imaged by the *HST* (shown on the right) and revealed seven white dwarfs (marked by circles). This picture clearly demonstrates the sensitivity of the *HST* because even the brightest white dwarf in the picture is no more luminous than a 100-watt light bulb seen at the Moon's distance (384 400 km or 239 000 miles) (credit: H. Richer and NASA).

The continued process of contraction also makes white dwarfs extremely dense. White dwarfs, which can weigh as much as the Sun, have sizes only comparable to that of the Earth. Their high density also means high gravity. The gravity on the surface of white dwarfs can be as high as a million times that on the surface of the Earth. In fact, if a piece of rock of mass 1 kg fell onto a white dwarf from outside, it would create as much energy as an earthquake measuring 7 on the Richter scale!

Shklovsky found many similarities between white dwarfs and central stars of planetary nebulae. They have similar temperatures and masses, with white dwarfs being less luminous and smaller. Since white dwarfs begins contracting when nuclear fuel is exhausted, is it possible that central stars of planetary nebulae are

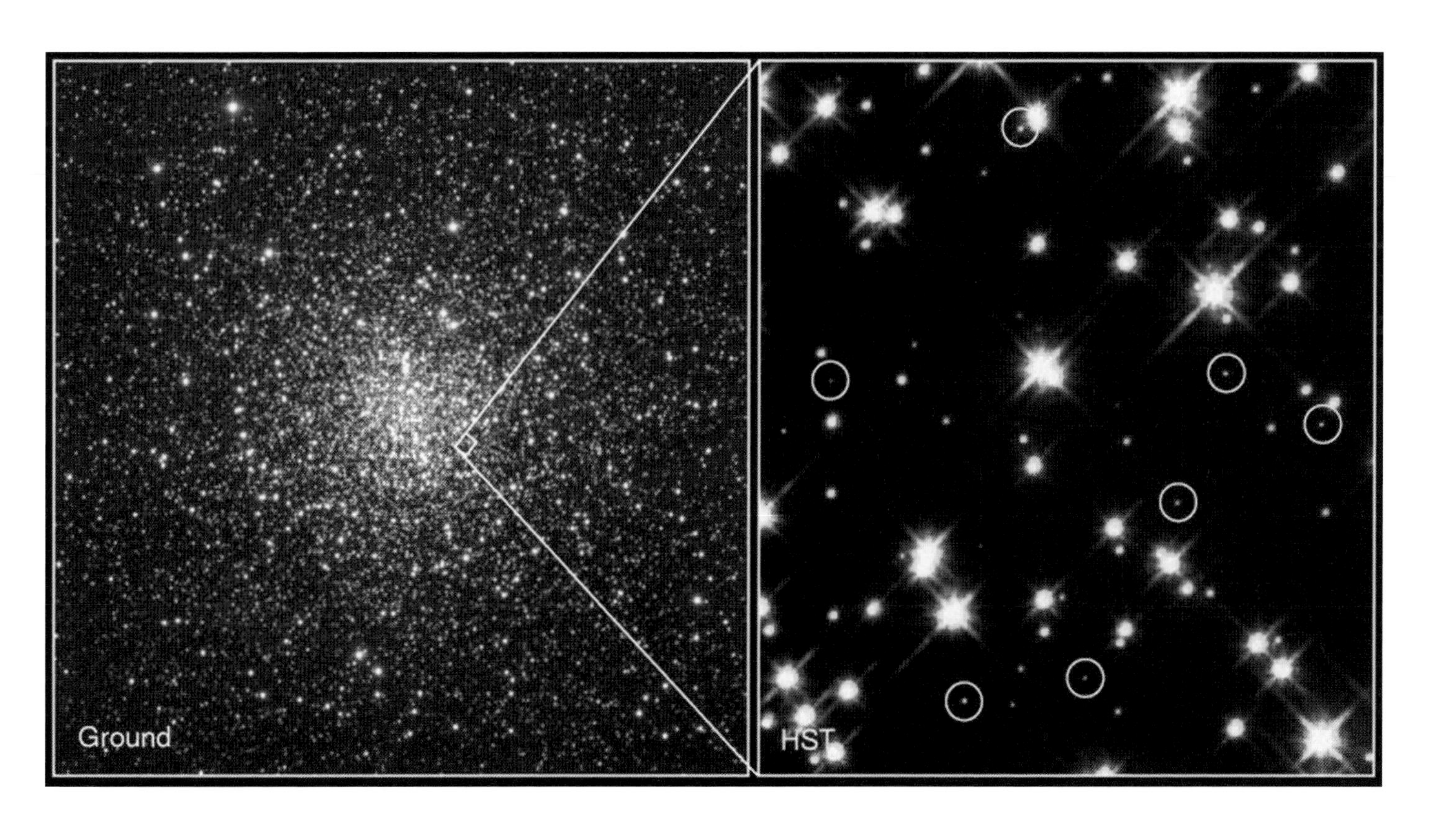

stars that are just about to run out of fuel? This led Shklovsky to hypothesize that white dwarfs evolve from central stars of planetary nebulae.

The picture is now complete. Planetary nebulae evolve from red giants and end up as white dwarfs. They therefore represent the last stage of stellar evolution just before the star fades away (see Fig. 5.1).

6 The end of a star's life: white dwarf, neutron star, or black hole?

Although stars are nothing but balls of gas, they manage to maintain a spherical shape and a stable size for billions of years. The stability is made possible by a delicate balance between the star's tendency to collapse by gravitational force and to expand by pressure from the hot gas inside. White dwarfs, however, have such high densities that the gravitational force is too strong to be resisted by gas pressure alone. Instead, a new force called 'electron degenerate pressure' takes over. In extremely high densities when electrons are squeezed against each other, they are forced to move at higher and higher speeds, according to the 'quantum statistical' theory of Fermi and Dirac. Just as the movement of molecules in the air exerts a pressure on the walls of a room, these fast moving electrons exert a pressure to resist gravitational attraction. This force is a purely quantum mechanical phenomenon and is not observable in our everyday lives. In 1926, R.H. Fowler of the University of Cambridge applied the just-published theory of Fermi and Dirac to the structure of stars. He found that the electron degenerate pressure could indeed support a star at high densities.

In July of 1930, Subrahmanyan Chandrasekhar was a 19-year-old on a voyage from India to England to study with Fowler. During the long boat journey, he realized that at the densities typical of white dwarfs, these electrons move near the speed of light. He then applied Einstein's theory of special relativity to the structure of white dwarfs and demonstrated that the maximum mass for white dwarfs is 1.4 solar masses, a value which has since been named the

Chandrasekhar limit. In recognition of this discovery, Chandrasekhar was awarded the Nobel Prize in 1983.

The existence of the Chandrasekhar limit suggests that a star born with a mass higher than 1.4 solar masses could not become a white dwarf. What could be the end state of such stars? In 1939, Robert Oppenheimer and George Volkoff of the University of California at Berkeley found that another stable state is theoretically possible. At densities higher than those in white dwarfs, electrons and protons can combine together to form neutrons. Neutrons have quantum properties similar to those of electrons, and an equivalent force called 'neutron degenerate pressure' can keep a star from collapsing. Such stars are called neutron stars. While a white dwarf with a mass of one solar mass is the size of the Earth, a neutron star of the same mass has a size of only 30 km. The actual existence of neutron stars remained in doubt for decades until the discovery of the pulsar in the summer of 1967 by Jocelyn Bell-Burnell of the University of Cambridge. A year later, Thomas Gold of Cornell University demonstrated that the pulsar phenomenon is a manifestation of rotating neutron stars.

In the early 1970s, it was commonly believed that the final state of a star at the end of its life is dependent on its mass. A star born with a mass of less than 1.4 solar masses will become a white dwarf, whereas a star two or three times the mass of the Sun will end as a neutron star. For stars more massive than that, there is no known stable state and they are assumed to collapse to an infinitesimal size. Such objects are called black holes because beyond certain densities, the gravitational attraction is so large that even light cannot escape. By definition, black holes cannot been seen and their existence can only be inferred by the gravitational effects on their stellar companions, or from the X-ray emissions that radiate as matter falls into a black hole.

We have seen that white dwarfs descend from central stars of planetary nebulae after the envelopes of their red giant progenitors are ejected. Similarly, neutron stars and black holes are the remnants of the cores of their progenitor stars after the stellar envelopes are ejected in the form of supernovae. If every star heavier than 1.4 solar masses becomes a supernova, then the creation of supernovae must be a very frequent occurrence. However, the last supernova seen in the Milky Way was discovered by Johannes Kepler in 1604. Even if we account for the supernovae on the other side of the Galaxy that are not detected because of dust extinction, the best estimate for the supernova rate in our Galaxy is no more than a few per century. There is clearly a discrepancy between theoretical expectations and observations. We will come back to the resolution of this problem in Chapter 9.

7 What is the source of power?

Although Shklovsky has provided a qualitative picture for the evolutionary status of planetary nebulae, there are many details that still need to be worked out. For example, how is the nebular shell ejected? What is the source of energy of the central star? If it generates energy by nuclear burning, what kinds of nuclear reactions are taking place? Shklovsky has already convinced everyone that planetary nebulae are evolved stars, so they probably have used up the supply of hydrogen in their cores. The other possibilities are helium and carbon burning. By nuclear reactions, helium (the product of hydrogen burning) can be converted into carbon, and carbon can be converted into oxygen, all releasing energy in the process. Using the new data from nuclear physics and mathematical models of stellar structure, theorists constructed elaborate models to try to explain why planetary nebulae are so luminous.

However, all of these models ran into a major hurdle. We know that red giants are red and central stars of planetary nebulae are blue (see Chapter 3). However, the lifetime of planetary nebulae is very short (only about 30 000 years), and the youngest planetary nebulae must have left the red giant stage no more than a few thousand years ago. The theories therefore had to explain how the star goes from about 3 000 °C (the temperature of red giants) to ten times as hot in such a short time as a few thousand years.

The struggles of the theorists were made more difficult by the observers. Astronomers usually test theories of stellar evolution by comparing their model results with the distribution of stars on a diagram plotting the stellar luminosity against the temperature. This diagram, known as the Herztsprung–Russell diagram, has its origins in the magnitude–color diagrams derived from star clusters. Astronomical magnitudes are a measure of stellar brightness, and colors are related to temperature. In order to have a valid test of the theory, we need an accurate plot of stars on the Hertzsprung–Russell diagram. This is not particularly difficult for normal stars on the main sequence but poses a great challenge to observers who wish to create this diagram for planetary nebulae. In order to convert the apparent brightness to intrinsic brightness, one needs to know the distances of planetary nebulae from us, which are highly uncertain (see Chapter 16). The central stars of planetary nebulae are often obscured by the nebular emission, making a measurement of their brightness and colors difficult. Furthermore, these stars emit a lot of energy in the ultraviolet that cannot be measured by ground-based observations.

In the mid-1960s, observers managed to come up with a Hertzsprung–Russell diagram for planetary nebulae. They suggested that the central stars of planetary nebulae begin their evolution at about 100 times the solar luminosity. As they get hotter, they also increase in brightness. After their brightness peaks at about 10 000 solar luminosities at a temperature of 50 000 °C, they then decrease in brightness as the stars continue to evolve to higher temperatures. At about 100 000 °C, they are back at 100 times the luminosity of the Sun. This observed evolutionary sequence for planetary nebulae is referred to as the Harman–Seaton Sequence and was widely used in textbooks for 30 years.

The theorists had a hard time trying to reproduce this diagram. They tried all kinds of tricks, but none of their fancy schemes came close to resembling the Harman–Seaton Sequence. It is simply difficult to make the star change temperatures and luminosities on such a short time scale. People were getting discouraged. Could Nature be so clandestine that we could not figure out the workings of planetary nebulae?

The breakthrough came in 1970. Bohdan Paczyński, a Polish astronomer working at the Copernicus Astronomical Center in Warsaw and an expert in stellar evolution theory, came to the conclusion that because of their high luminosities, planetary nebulae cannot be descendants of ordinary red giants. Their immediate progenitors must be 'asymptotic giant branch' stars, stars that are even redder and more luminous than regular red giants. At the end of the red giant stage, stars ignite helium in the core. After helium is burnt out, their ashes (carbon and oxygen) are deposited in an inert core. This marks the beginning of the asymptotic giant branch. Above the core are layers of hydrogen and helium which continue to burn. Further out is a huge envelope

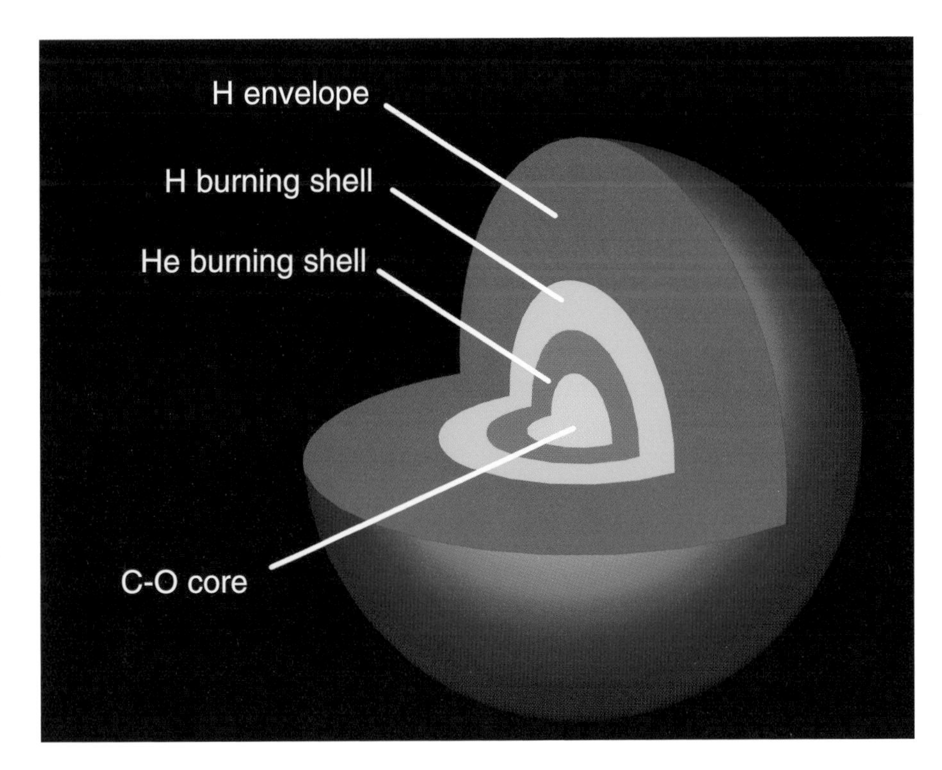

Figure 7.1.
A schematic diagram illustrating the structure of an asymptotic giant branch star. At the center is a core made up of oxygen and carbon (C–O). Nuclear fusion takes place in two separate layers, an inner layer burning helium into carbon and an outer layer burning hydrogen into helium. An outer hydrogen envelope occupies most of the volume of the star. Asymptotic giant branch stars are among the largest stars known. When the Sun becomes an asymptotic giant branch star, it will likely expand to engulf all the inner planets, including Mars. This drawing is not to scale. The actual sizes of the core and the two burning shells are much smaller than illustrated here.

made of hydrogen (Fig. 7.1). As an asymptotic giant branch star evolves, it becomes larger and more luminous. A star near the end of its asymptotic giant branch evolution can be ten thousand times brighter than the Sun, with a size two to three times that of the Earth's orbit. Asymptotic giant branch stars are some of the brightest stars in the Galaxy.

If planetary nebulae evolve from asymptotic giant branch stars, then their temperatures have to increase by a factor of ten in a few thousand years. Asymptotic giant branch stars are red because of their huge hydrogen envelopes. If the envelope is peeled off, we should see the hot core and the star should appear bluer. Through some simple calculations, Paczyński was able to show that even a very small amount of hydrogen in the envelope is sufficient to prevent the star from changing color (i.e. to move from the right side of the Hertzsprung–Russell diagram, where the asymptotic giant branch stars reside, to the left side, where the planetary nebulae are found). Suppose that we can remove the entire envelope, by ejection or some other means, to reveal the core, stripping the envelope to less than 0.1% of its original mass. With such a small envelope, the star will evolve from red to blue at constant luminosity – in other words, flat across the Hertzsprung–Russell diagram (Fig. 7.2).

This is in total contradiction to the observed Harman–Seaton Sequence, which says that the stars change in brightness by a factor of 100 over several thousand years. Paczyński had so much confidence in his calculations that he defended his ideas vigorously against some skeptical observers during the 1976 International Astronomical Union symposium on planetary nebulae in Ithaca, New York. However, the tide was already turning in 1976. The observational errors in the derivation of luminosities and temperatures of central stars of planetary nebulae turned out to be larger than previously believed. It was conceivable that the observations were wrong and the Harman–Seaton Sequence was fictitious.

Figure 7.2.
[opposite] A schematic diagram showing the life cycle of a solar-like star. For about 10 billion years, a star similar to the Sun maintains its luminosity by burning hydrogen into helium in the core. After helium is exhausted in the core, the envelope of the Sun expands to become a red giant, while its core contracts. A red giant draws its energy from burning hydrogen in a layer above the core. After further contraction of the core, helium can be ignited in the core, followed by further expansion of the envelope. As the star enters the asymptotic giant branch, its core is made up of carbon and oxygen, ashes of previous nuclear burnings. Over a period of about one million years, the star expands even further, reaching a size as large as the orbit of Mars. The luminosity also increases to several thousand times the luminosity of the present Sun. A stellar wind gradually depletes the hydrogen envelope and exposes the hot core. The remnants of the stellar wind become a planetary nebula, and the exposed core becomes the central star of the nebula. As all the hydrogen of the central star is burned up, it dims and contracts to become a white dwarf.

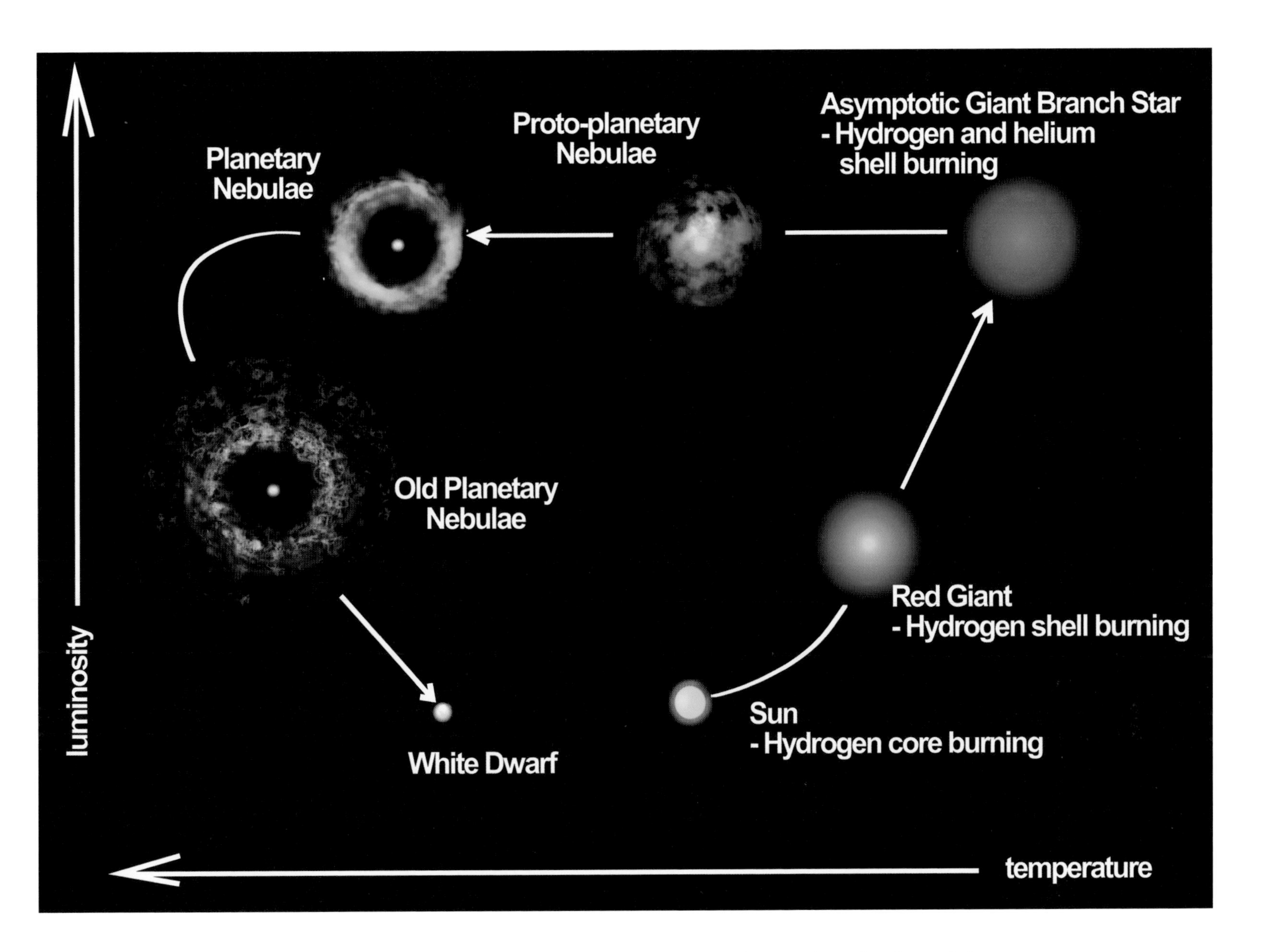

Subsequent independent calculations by Icko Iben of the University of Illinois and Detlef Schönberner of the University of Kiel have confirmed the conclusions of Paczyński. The central stars of planetary nebulae are burning hydrogen in a thin layer above a core made up of carbon and oxygen. When all the hydrogen is burned up, the star will fade and decrease in luminosity. This will take several tens of thousands of years, in agreement with the observed lifetimes of planetary nebulae.

8 Star dust

While the problem of nuclear processes was solved by Paczyński, the problem of how the envelope is ejected remained. The conventional wisdom in the early 1970s was that when asymptotic giant branch stars get too large in size, the envelope becomes unstable. This results in the envelope being detached from the core and becoming a planetary nebula. Qualitatively this seems to be a reasonable idea. After all, violent events are rampant in the Universe. In a supernova explosion, a huge amount of mass is thrown out of a star. Novae eject shells after a nova explosion. Why not asymptotic giant branch stars?

After Shklovsky's paper in 1956, a wide variety of theories were proposed on how to eject the red giant envelope. All these efforts have ended in failure. These theories either eject too much or too little mass, never the one tenth of a solar mass that is observed in the planetary nebulae shells. The situation got worse after the acceptance of Paczyński's theory. According to Paczyński, the star has to leave behind a very precise amount of matter after the ejection. The amount of precision required is between 0.001 % and 0.1% depending on the mass of the star. If the ejection leaves behind just a tiny bit more, this extra mass has to be burnt before the core is exposed. Without exposing the core, the star will stay red and not change color. At the same time, the ejected envelope is expanding and dispersing. This delay in evolving to high temperatures means no ultraviolet photons will be emitted to ionize the ejected envelope. Since ultraviolet photons are what make the nebula shine (see Chapter 3), there would be no planetary nebula. The ejected envelope would flow away from the star without ever being noticed.

No violent ejection can be that precise. The theories had problems even ejecting the right amount by a factor of 10. It was asking far too much for these theories to be precise within 0.1%! It seemed that we were stuck again. Before we come back to the resolution of this problem, we need to mention a development in infrared astronomy.

In the late 1960s, Ed Ney, a solid state physicist at the University of Minnesota, had just built an infrared detector which was capable of detecting infrared radiation with wavelengths as long as 10 μm. He put this detector on a 30-inch telescope near St Croix river just outside of Minneapolis. Together with Nick Woolf, a professor of astronomy in Minnesota, Ney began to observe stars with this new instrument.

It has been known since the end of the nineteenth century that the color of objects can change with temperature. When we put an iron rod in the fireplace, it becomes red in color. As it stays in the fire longer and its temperature increases, the color will gradually change to orange to whitish. This phenomenon is known as Wien's law, which says that a body of high temperature will radiate a shorter wavelength (Fig. 8.1). It is through the application of Wien's law that we know that a red star has lower temperature than a blue star.

According to Wien's law, a red giant of 3000 °C would have a wavelength with maximum brightness of 1 μm. Since red giants have the lowest temperature of all known stars, there was very little expectation that observing at wavelengths longer than 1 μm would yield anything interesting. This was not to be the case. With their infrared telescope, Woolf and Ney were finding stars that are extraordinarily bright at the wavelength of 10 μm. Red giants in particular are much brighter at 10 μm than expected. Woolf and Ney suggested that this excess is due to dust surrounding the star.

Figure 8.1.
A graphical illustration of Wien's law, the relationship between color and temperature. The three curves represent the energy emitted by radiators of three different temperatures. The peak of the radiation shifts to shorter wavelengths as the temperature of the object increases. For example, the Sun has a temperature of about 6000 °C and appears yellow, whereas a star of lower temperature will appear red in color.

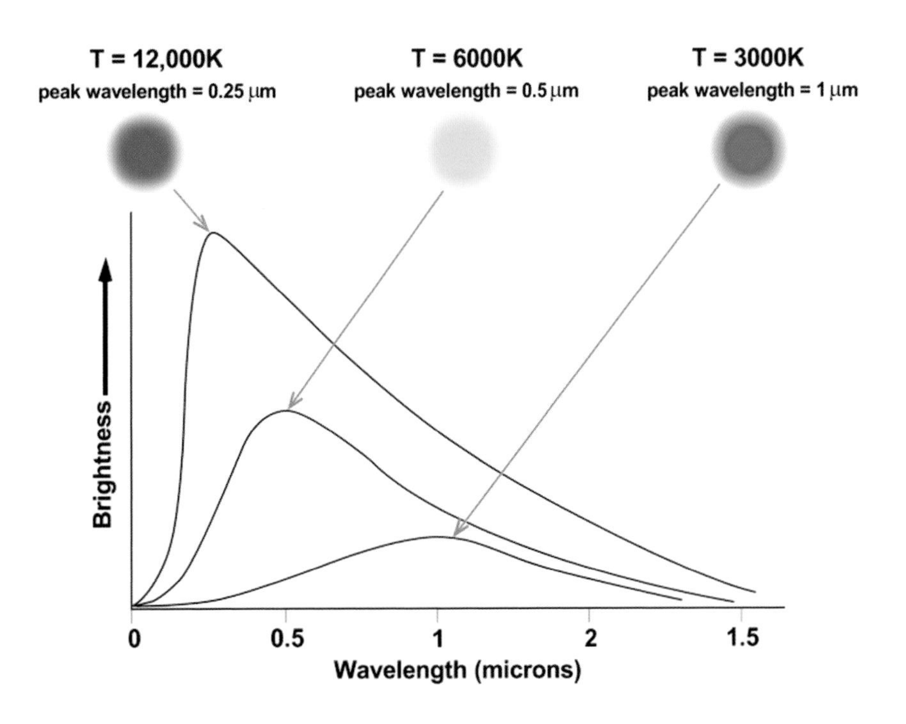

The star dust that Woolf and Ney saw is not that different from the ordinary dust in our Earth environment. In addition to molecules (mostly nitrogen and oxygen), the Earth's atmosphere contains many other particles too small to be seen by the naked eye. In heavy concentrations, however, their presence can be detected because they absorb and reflect sunlight. For instance, dust particles created by wood burning stoves can be seen as smoke coming out of chimneys. The magnificent red sunsets that we see are the result of the blue light of the Sun being deflected by dust in the atmosphere. Less familiar to us is the infrared radiation that is given off by dust particles. All objects at room temperature (our bodies, trees, rocks, dust) radiate in the infrared. Since our eyes are not responsive to infrared radiation, we are not aware that things shine in the infrared even in a dark room. However, if equipped with an infrared sensor, we can see that everything glows in the dark.

The infrared excess seen in stars is explained by circumstellar dust absorbing visible radiation from the star and re-emitting it in the infrared. Everything on Earth – plants, animals, our bodies – absorbs sunlight when exposed to the Sun. Such energy does not disappear. Plants use some of the solar energy for photosynthesis. Objects like stones reflect back some of the sunlight and absorb the rest. The absorbed energy allows the object to become warm and to self radiate. The re-radiation occurs at longer wavelengths than the received wavelength; for room temperature objects, it occurs in the infrared. So even at night or in a dark room, our surroundings are filled with infrared radiation coming from living objects like our bodies and non-living things such as chairs and tables. Since the Sun is the dominant source of radiation only in the visible but not in the infrared, if we had infrared eyes there would be little difference between day and night.

Astronomers knew about the existence of dust in the interstellar space because dust particles absorb starlight. Interstellar dust can shine in the infrared if it is warm. Such is the case near a star. The Minnesota observations suggest that red giants are capable of producing dust in their atmosphere. This dust is heated by the starlight and becomes warm, making it visible to infrared telescopes. By observing at several wavelengths near the atmospheric window at 10 μm, Woolf and Ney were able to identify the composition of the dust. The dust is made up of silicates, similar to beach sand although much smaller.

More importantly, they found that the amount of dust is highest among asymptotic giant branch stars (see Fig. 8.2). In some cases, there is so much dust that the star itself is totally obscured. The most dramatic example is the star CW Leo. In the visible, it is an innocuous star of 19th magnitude. But in the infrared, it is the one of the brightest stars in the sky! The visible faintness, and the corresponding infrared brightness, is due to circumstellar dust.

These discoveries suggest that there could exist many stars in

Figure 8.2.
[opposite] A schematic diagram showing the circumstellar envelope of dust and gas built up by the stellar wind from an asymptotic giant branch star. Over a period of several hundred thousand years, the envelope can be as large as a couple of hundred light years in size.

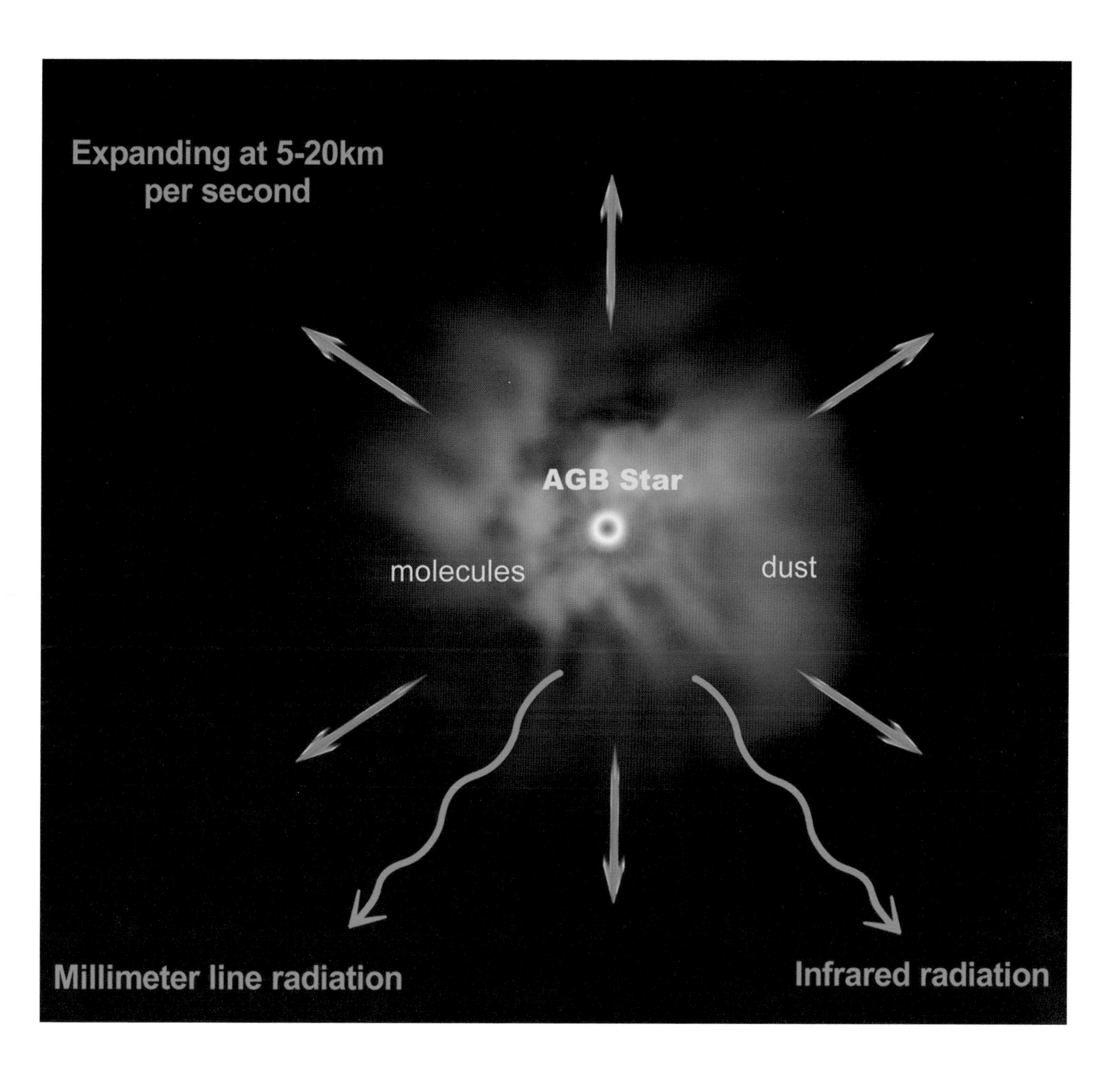
Expanding at 5-20km
per second
AGB Star
molecules
dust
Millimeter line radiation
Infrared radiation

the sky that we don't know about. The visible sky does not give a complete picture.

Asymptotic giant branch stars are not alone in having dusty envelopes. Strong infrared radiation attributed to dust was detected in the planetary nebula NGC 7027 in 1967. Planetary nebulae are dusty places.

9 Gone with the wind

In 1971, Philip Solomon, then at the University of Minnesota, used the newly developed millimeter-wave receiver by Bell Labs to observe CW Leo with the 12-m *National Radio Astronomy Observatory* radio telescope at Kitt Peak, Arizona. He found a very strong signal from the star due to the molecule carbon monoxide (CO). This detection suggested that there are large amounts of gaseous molecules mixed with the dust. These molecules must have been synthesized in the atmosphere of the star and later ejected by the star. Observations also tell us that the molecules are flowing out at a speed of about 15 km/s. From the strength of the signal one can estimate that the star is losing mass at a rate of one hundred-thousandth of a solar mass per year (or 600 trillion tonnes per second). This is a very high rate because if it is kept up, a solar mass is lost in only one hundred thousand years. In comparison, the Sun releases only one-hundred-trillionth, or 10^{-14}, of its mass per year. Further millimeter-wave observations showed that CW Leo is not an isolated case. Jill Knapp of Princeton University found that almost every asymptotic giant branch star showing large infrared excess also shows CO emission. This demonstrated without a doubt that large-scale mass loss is a common occurrence during the asymptotic giant branch phase.

The discovery of mass loss from asymptotic giant branch stars has significant implications for how stars end their lives. Since the asymptotic giant branch lasts approximately one million years, if such high rates apply throughout the lifetime of the asymptotic giant branch, then a significant amount of mass could be lost. However, news travels slowly in the astronomical community and most stellar evolutionary models computed in the 1970s made little or no allowance for mass loss. In a conference on red giants held in the hilltop town of Erice on the island of Sicily in 1980, observers and theorists finally got together and debated the proper mass loss formula for stellar evolutionary models. There was

Figure 9.1.
The open cluster Pleiades in the constellation Taurus. A white dwarf discovered in this cluster provided the first convincing proofs that most stars become white dwarfs instead of neutron stars or black holes. Photograph by Wei Hao Wang.

a lot at stake because if a high mass loss formula was adopted, then a star could lose several solar masses during the asymptotic giant branch phase, and a 3, 5, or even 8 solar mass star could become a white dwarf!

There were already some hints that this is indeed the case. In 1960, Willem Luyten of the University of Minnesota and George Herbig of *Lick Observatory* found a white dwarf in the Pleiades cluster (Fig. 9.1). Since all stars in a cluster are supposed to be born at the same time and since high mass stars evolve much faster than low mass stars, the existence of a white dwarf in a cluster implies that this white dwarf must have evolved from a relatively high

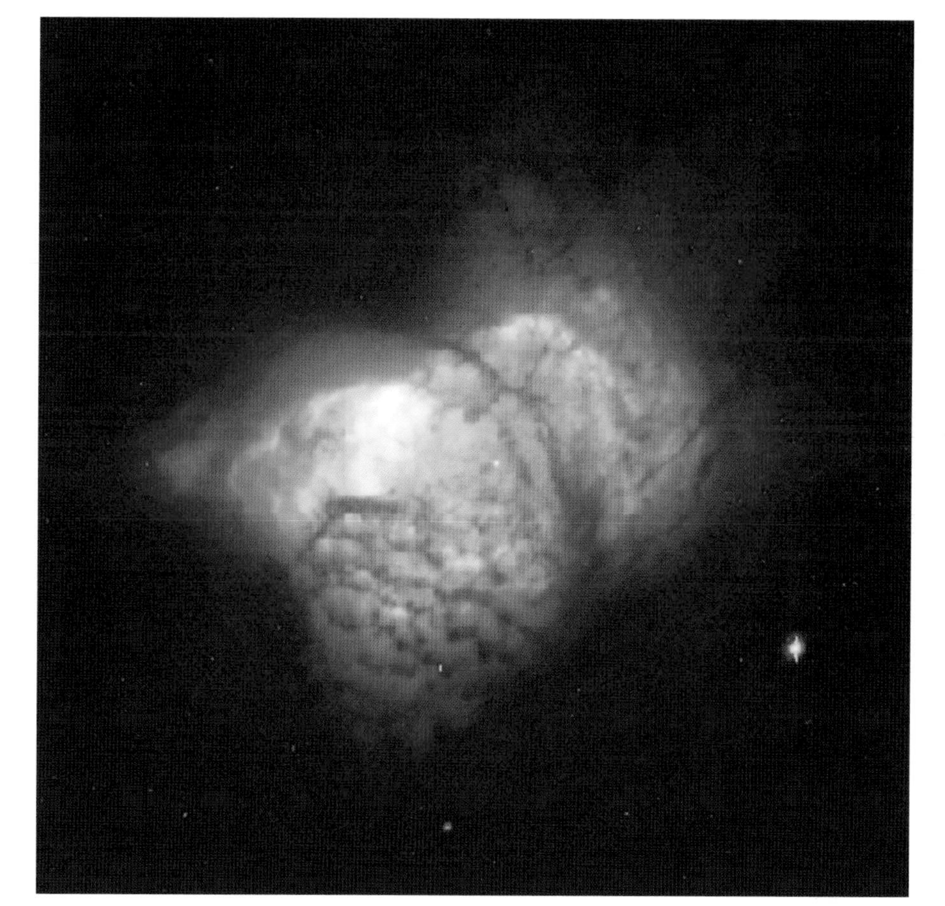

Figure 9.2.
This small (10 arcsec) nebula in the constellation of Cygnus has a bright and strong emission-line spectrum and is the most extensively observed among all planetary nebulae. The dark lane that wraps around the central part of the nebula is due to dust absorption. The temperature of the central star of the planetary nebula NGC 7027 is approximately 200 000 °C, making it one of the hottest stars known. Comparison between this observed temperature with the theoretical stellar evolutionary models of Schönberner suggests that the central star has a current mass of about 0.7 solar masses and that it left the asymptotic giant branch about 700 years ago. This stellar mass, together with the 3 solar masses found in the molecular envelope (see Chapter 10), implies that it must have descended from a star with an initial mass of at least 4 solar masses. NGC 7027 therefore provides further evidence that stars that are born with masses several times higher than the Sun will not become supernovae, but end their lives as planetary nebulae instead (credit: Robin Ciardullo).

mass star. In the case of Pleiades, which is about 100 million years old, this white dwarf must have descended from a star with an original mass at least six times that of the Sun.

Mass loss on the asymptotic giant branch therefore became a convenient, or perhaps even necessary, explanation for the existence of this white dwarf. More recent observations of white dwarfs in galactic clusters by Dieter Reimers of the University of Hamburg put the upper mass limit for the formation of white dwarfs at 8 solar masses. This means that all stars with initial masses of less than 8 solar masses will become white dwarfs instead of neutron stars or black holes. Taking into account the fact that most of the stars are born with low masses, it is estimated that 95% of stars will end their lives as white dwarfs.

While stars maintain a constant mass through most of their billion-year lifetime, they undergo a rapid weight loss in the last one million years of their lives. This last minute weight loss allows a star to avoid a violent death as a supernova. In spite of the glamorous status of supernovae and black holes in the minds of many people, they are relevant to the lives of only a small fraction of all stars. Planetary nebulae, not supernovae, are the fate of most stars. Unlike old soldiers, old stars do die. But, like them, most just quietly fade away.

10 Not with a bang but a whimper

If planetary nebulae descend from stars with initial mass as high as 8 solar masses, where is this mass now? The problem of missing mass got me thinking about the connection between asymptotic giant branch mass loss and planetary nebulae ejection. The amount of mass seen in planetary nebulae (a fraction of a solar mass) is small compared to the amount of mass ejected over the lifetime of the asymptotic giant branch. If there are already several solar masses of material in the immediate surrounding of a star at the end of its asymptotic giant branch evolution, then such material must have an effect on the subsequent planetary nebulae phase.

The question is: how? The circumstellar envelopes of asymptotic giant branch stars are hollow and round, whereas planetary nebulae have ring shapes and much higher densities. Clearly, it is not just a question of these circumstellar envelopes gradually diffusing into the interstellar medium and being seen as planetary nebulae. Furthermore, planetary nebulae are known to expand about twice as fast as winds from asymptotic giant branch stars. If planetary nebulae originate from these circumstellar envelopes, then some other mechanism is needed to compress and accelerate the nebula.

In the 1970s, Chris Purton of York University was hosting gatherings at his farm in Erin, Ontario, which he affectionately called the Erin Institute of Astrophysics. At the 1977 meeting, Chris Purton, Pim FitzGerald (of University of Waterloo) and I came up with a theory for planetary nebulae formation which is now commonly referred to as the interacting winds theory. The circumstellar envelopes of asymptotic giant branch stars are built up over a period of several hundred thousand years through a gradual process of a slow stellar wind. If this wind is able to completely deplete the atmosphere of an asymptotic giant branch star, then the stellar core will be exposed as a hot star. A separate stellar wind was hypothesized to emerge from this hot star at high speed, acting as a snow plow and sweeping up the slow wind material left over from the asymptotic giant branch. The swept up material makes a dense shell, which is what we traditionally call the planetary nebula.

In 1978, we published the dynamical calculations of the interacting winds theory in the *Astrophysical Journal*. Using parameters of observed asymptotic giant branch stars, we estimated

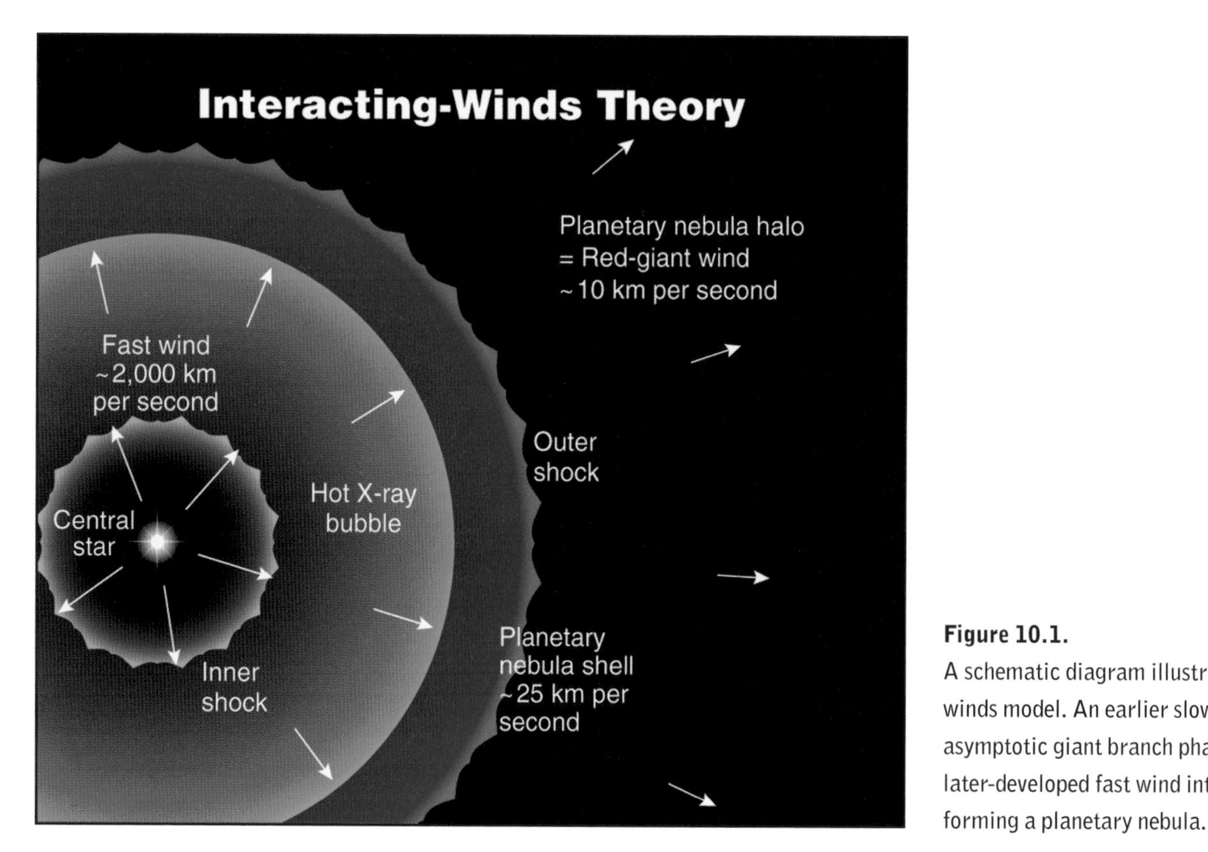

Figure 10.1.
A schematic diagram illustrating the interacting winds model. An earlier slow wind from the asymptotic giant branch phase is swept up by a later-developed fast wind into a dense shell, forming a planetary nebula.

that the hypothetical fast wind needs to be moving at thousands of kilometers per second in order to produce the observed density and expansion velocities of planetary nebulae. Although this seemed to be an incredibly fast wind, we thought that the central stars of planetary nebulae, because of their high luminosity and high temperature, could exert enough light pressure on atoms and ions to accelerate them to such speeds.

In the interacting winds model, as illustrated in the schematic diagram in Fig. 10.1, planetary nebulae have three mass components: the fast wind, the swept up shell, and the remnant of the slow wind outside the shell. We have to keep in mind that the term 'slow' is relative, because even this slow wind is moving at 10–20 km/s, or 36 000–72 000 km/h! Because the outer material has a much lower density and is outshone by the bright dense inner nebula, it is difficult to detect.

Observational confirmation of the theory came very quickly. On January 26, 1978, the *International Ultraviolet Explorer* (*IUE*) satellite was launched, carrying the first major ultraviolet telescope capable of spectroscopic measurements (Fig. 10.2). One of the first discoveries that came out from *IUE*, made by Sara Heap of Goddard Space Flight Center, was the detection of fast stellar winds from central stars of planetary nebulae. These winds are found by *IUE* to have remarkable speeds of 2000–4000 km/s, and therefore are among the fastest moving phenomena observed in the Universe (see Fig. 10.3). Further observations by Mario Perinotto of *Observatory of Florence* confirmed that fast winds from central stars are very common, and there is no doubt that such winds would have significant effects on the dynamics of the nebula.

With the 1980s came the CCD revolution. CCDs (charged coupled devices) are digital devices that replaced photographic plates as the main method to take astronomical images in the visible. These are the same devices that make up the view finders of home video cameras. The high dynamic range of CCD detectors

Figure 10.2.
The *International Ultraviolet Explorer* (*IUE*) was a joint project between NASA, ESA, and the United Kingdom. It was the first astronomical observatory to be placed in high Earth orbit. Launched on January 26, 1978, from Kennedy Space Center into a geosynchronous orbit over the Atlantic Ocean, it allowed constant communication with the ground stations. This picture illustrates how the *IUE* communicates with its two ground stations in Greenbelt, Maryland (16 hours a day) and Villafranca, Spain (8 hours a day). Astronomers observing at the ground stations were able to send real-time commands to the telescope, and receive and monitor the data as they came in. Since *IUE* observers are able to inspect their data and change their observing program according to the results, it operated very much like a ground-based telescope. The *IUE* carried a 45-cm telescope and was in operation for almost 20 years. It was one of the most successful and productive space astronomy telescopes ever built (credit: NASA).

makes it possible to measure faint objects in the vicinity of bright ones. You-Hua Chu (University of Illinois) and George Jacoby (*Kitt Peak National Observatory*) made use of this new technology and discovered that many planetary nebulae have faint halos. Although the existence of halos around some nebulae was already known since the 1930s, CCD observations brought home the message that they are extremely common. Bruce Balick of the University of Washington, working with Jacoby, took some very spectacular pictures of planetary nebulae with halos using the 4-m Mayall telescope of *Kitt Peak National Observatory*. In Figs. 10.4 and 10.5, we show two examples of faint halos observed by Balick and Jacoby around the famous planetary nebulae NGC 6826 and NGC 6543. The common presence of halos, due to the remnant of the circumstellar envelope, confirms the prediction of the interacting winds theory that the shells of planetary nebulae are swept-up material of the asymptotic giant branch wind. Several examples of halos are shown in Figs. 10.6–10.12.

Figure 10.3.
[opposite] The central star of NGC 6751 in the constellation Aquila is among the many planetary nebulae central stars to have been found to have a fast stellar wind by the *IUE*. An analysis of the ultraviolet spectrum of NGC 6751 by Walter Feibelman of Goddard Space Flight Center shows that the stellar wind has a speed of 2500 km/s. For comparison, a fast car can go 200 km/h, or 0.05 km/s. A bullet travels at about 1 km/s. These winds are very fast indeed! (Credit: A. Hajian.)

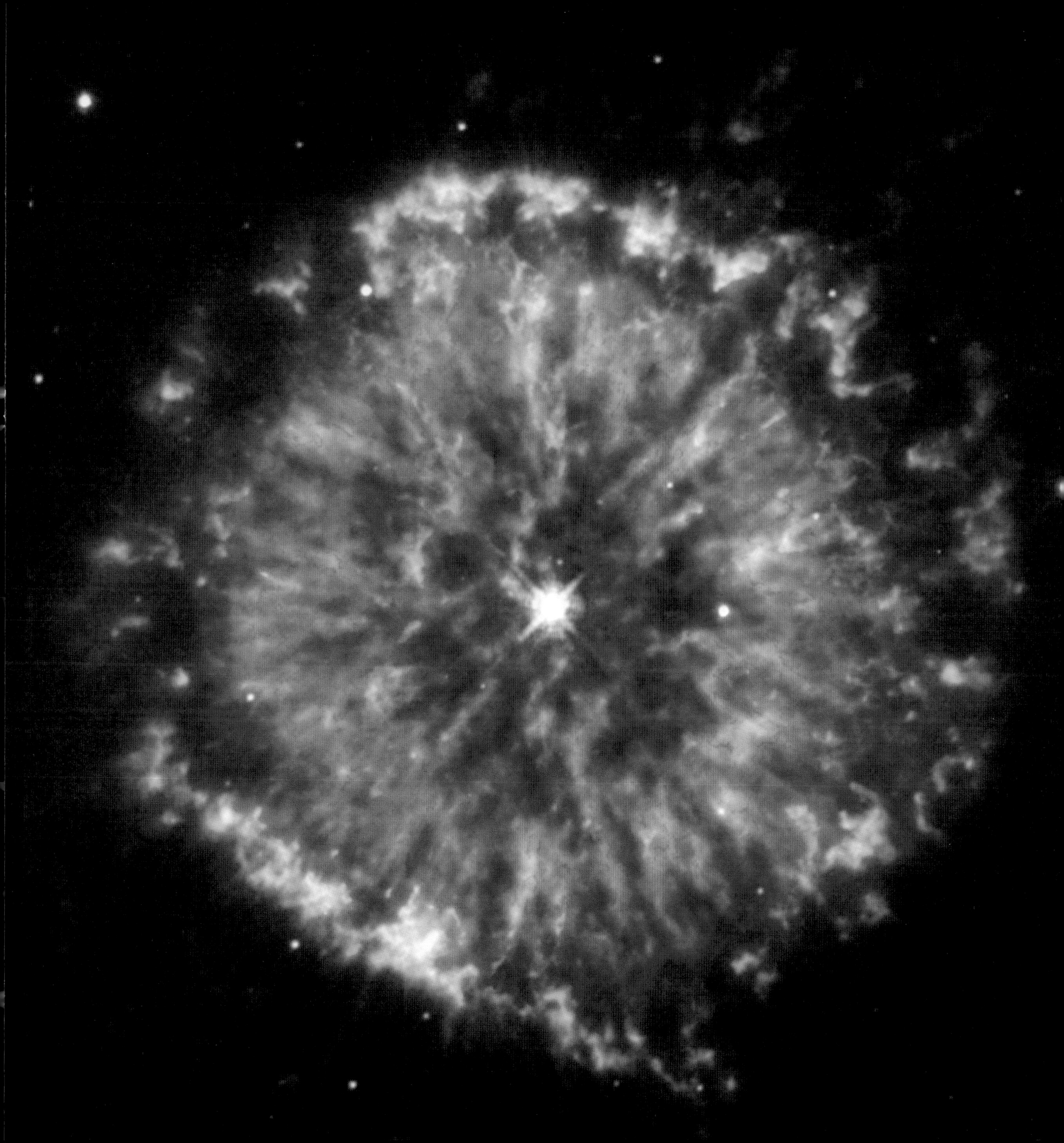

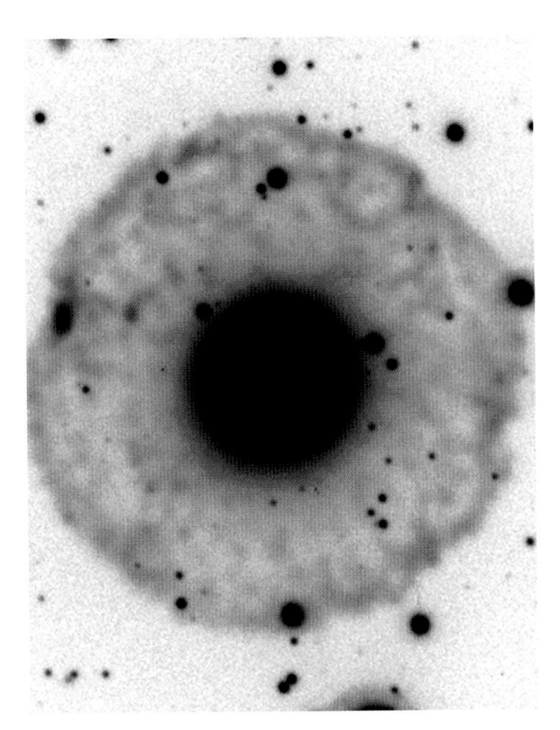

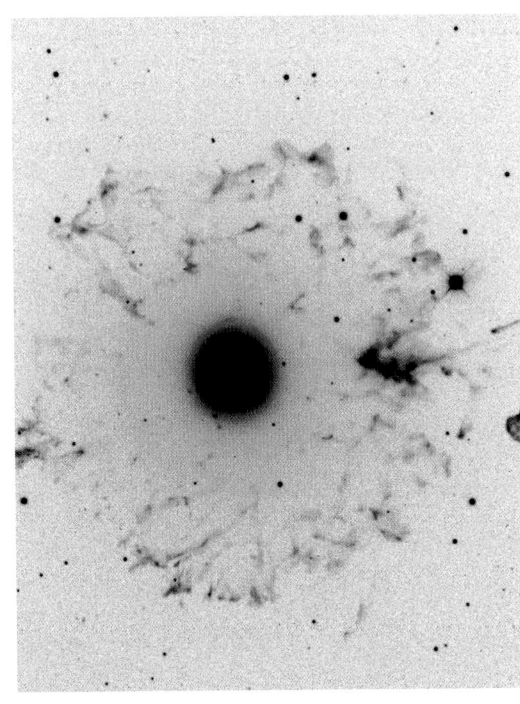

Figure 10.4.
[left] A deep image of NGC 6826 taken by Bruce Balick and George Jacoby at *Kitt Peak*. A large halo is detected outside the main nebula, which is overexposed in order to reveal the halo. An *HST* image of the nebula is shown in Fig. 15.1 (credit: George Jacoby and NOAO).

Figure 10.5.
[right] The Cat's Eye Nebula is surrounded by a large, patchy, filamentary halo of 5 arcmin in size. The core of the nebula (shown in Fig. 2.8) is only one-tenth the size of the halo. In order to reveal the faint halo, the core is totally saturated in this *KPNO* 4-m image (credit: George Jacoby and NOAO).

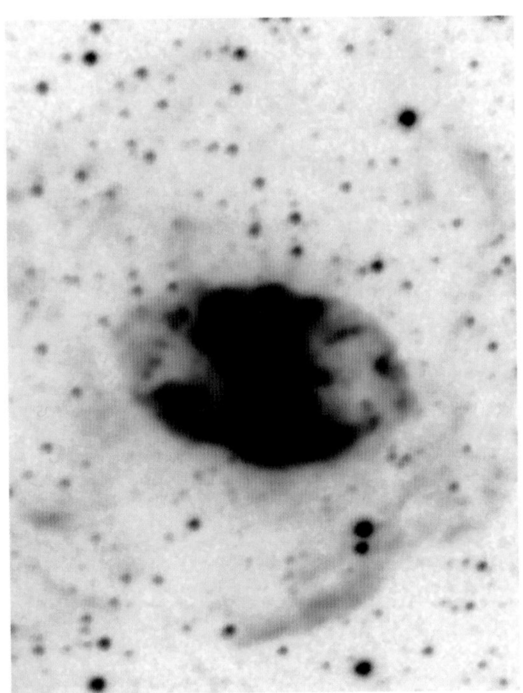

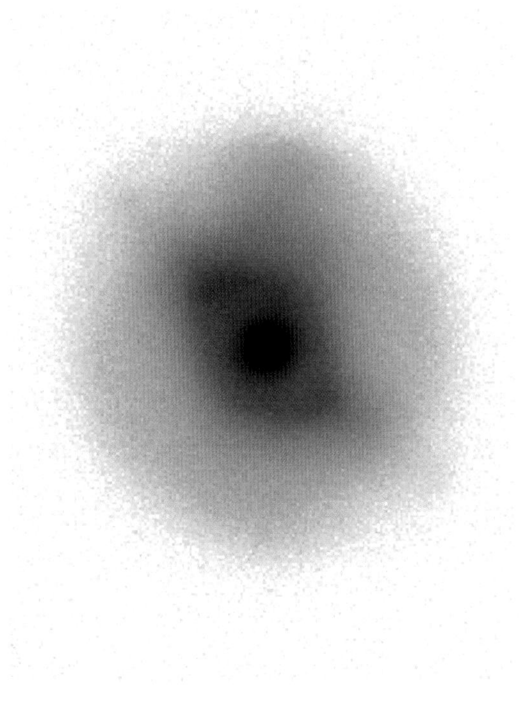

Figure 10.6.
[left] This ground-based picture of He 2-119 by Trung Hua shows the clear presence of an outer halo (credit: Hua).

Figure 10.7.
[right] When a deep exposure is taken on NGC 6891, a spherical halo of 80 arcsec in diameter is revealed. The structure of the main inner shell, shown as an overexposed core in this image, can be seen in Fig. 2.1. This picture was taken by Luis Miranda and Martin Guerrero with the 2.56-m *Nordic Optical Telescope* in La Palma, Spain.

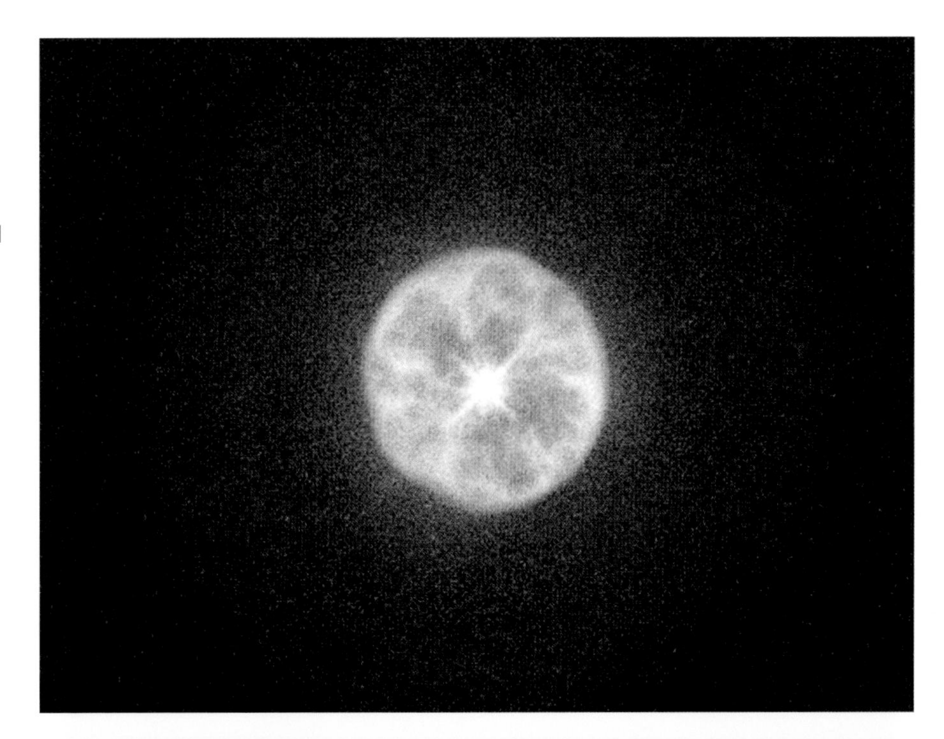

Figure 10.8.
Faint halos are commonly found outside the shells of planetary nebulae, as in this *HST* picture of IC 3568. The halo (about 20 arcsec in diameter), is approximately twice the size of the inner shell. IC 3568 is one of the most circularly symmetric of all planetary nebulae. In the words of Bruce Balick, 'if IC 3568 did not exist, it might have been created by theoreticians'. Its simple geometry allows easy comparison with dynamical models based on the interacting winds theory.

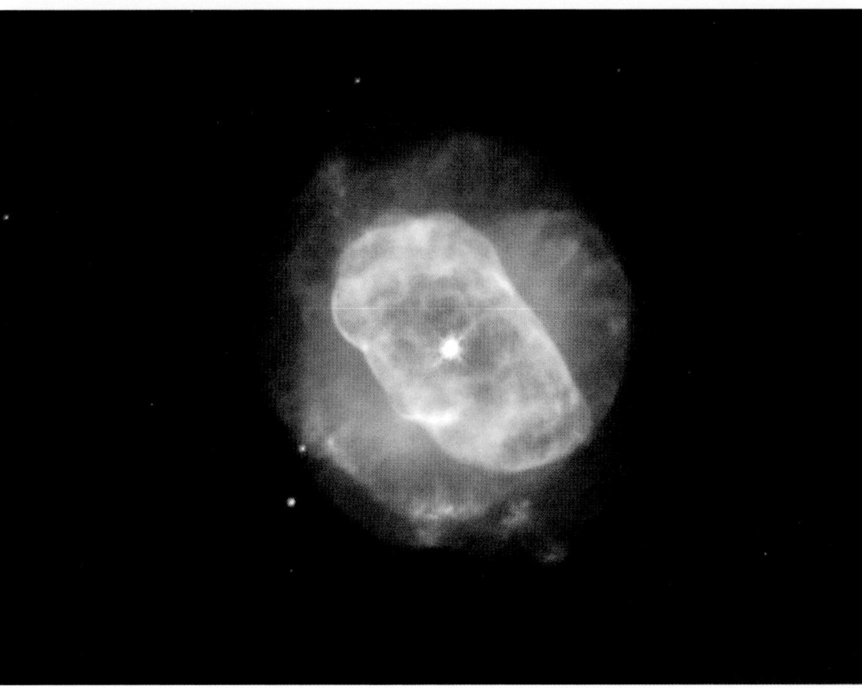

Figure 10.9.
The bright southern planetary nebula NGC 5882 was found to have a faint halo by Bill Weller and Steve Heathcote of *Cerro Tololo InterAmerican Observatory*. The halo has a size of 90 arcsec, extending far beyond the structures shown in this *HST* WFPC2 image. Such a multiple shell structure is not uncommon among planetary nebulae and at first seemed very difficult to explain. Detailed computer simulations based on the interacting winds theory, however, successfully reproduce such structures.

Figure 10.10.
[opposite] NGC 5979 is another example of planetary nebulae with multiple shells. The fainter structure outside of the main nebular shell is referred to as 'crowns'. Beyond the 'crown' is the halo.

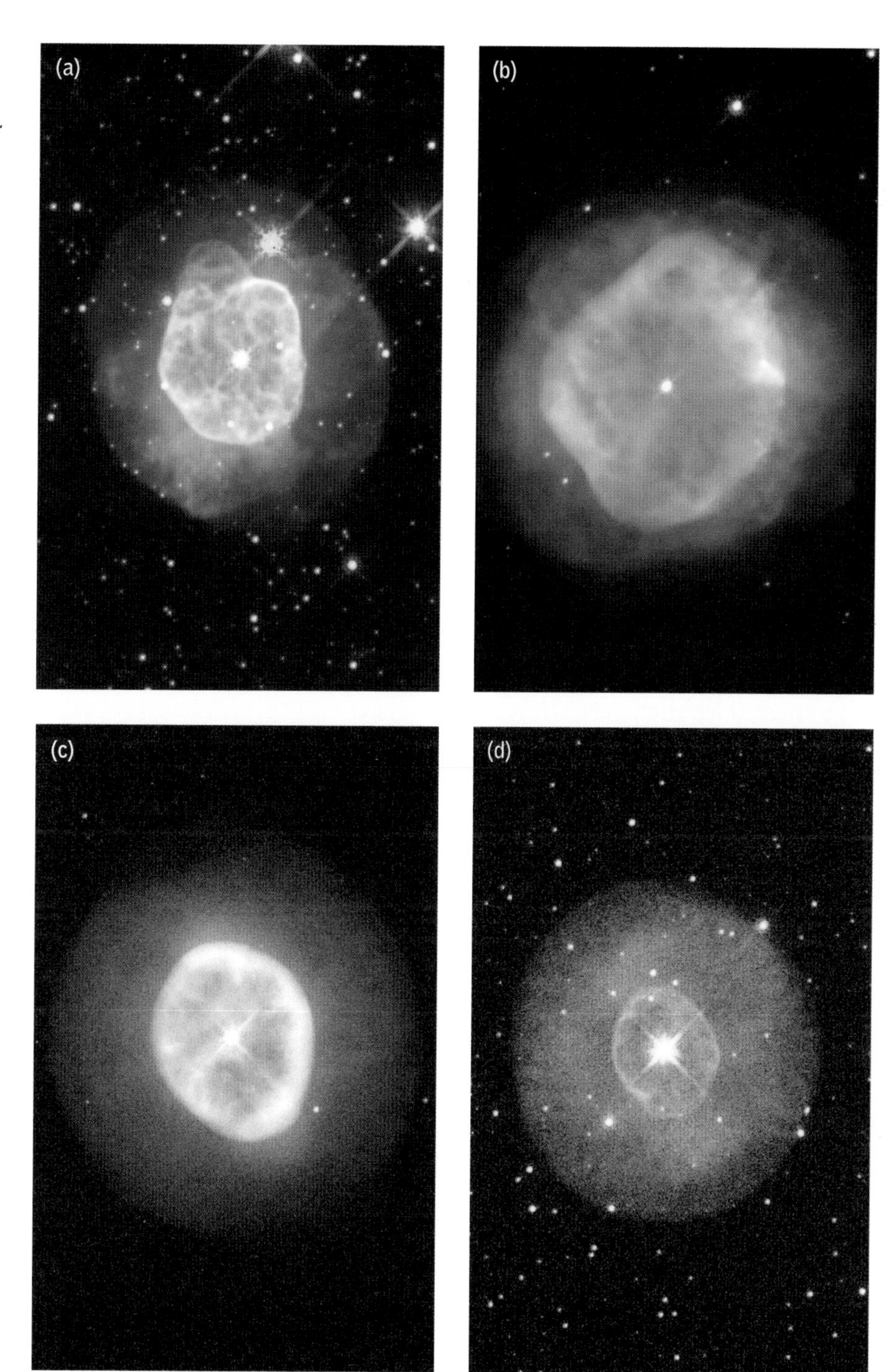

Figure 10.11.
Although halos and 'crowns' were discovered in many planetary nebulae by ground-based observations using CCD detectors, the most spectacular pictures illustrating these phenomena are provided by the *HST*. Here we show the 'crowns' of the planetary nebulae (a) NGC 6578, (b) NGC 2792, (c) IC 2448 and (d) NGC 6629 as imaged by the *HST*.

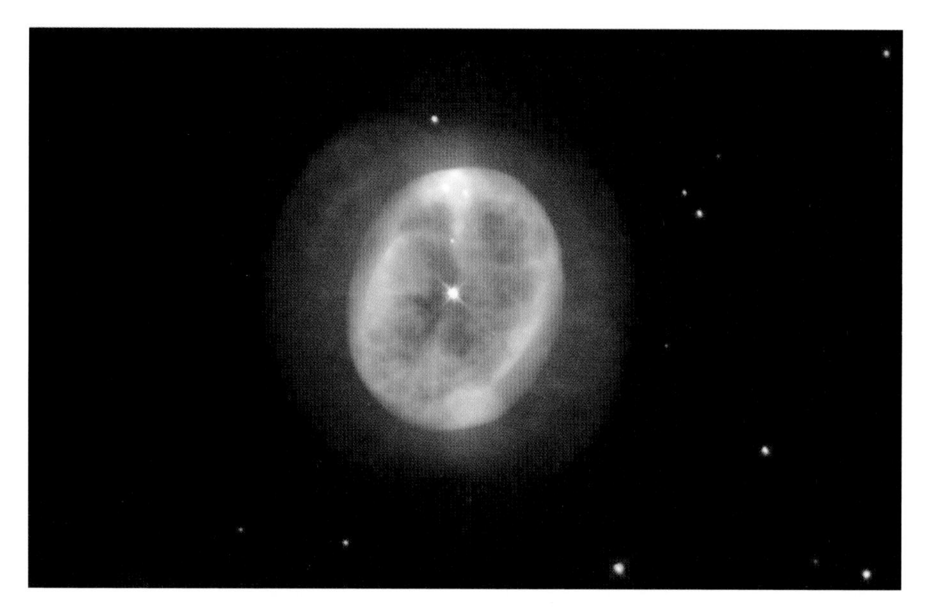

Figure 10.12.
This *HST* WFPC2 image of NGC 2022 shows that outside of the elliptical ring (22"×17") is a faint outer 'crown' about half an arcminute in diameter.

If planetary nebulae are just swept up wind material, then the dust ejected by the parent asymptotic giant branch stars should still be present in planetary nebulae. In order to confirm the asymptotic giant branch–planetary nebulae connection, it would be important to find the remnants of such dust in planetary nebulae. However, over the period of several thousand years since the star left the asymptotic giant branch, the dust has expanded away from the star. The increased distance from the central star makes the dust much cooler, maybe as low as −200 °C. According to Wien's law, the peak of dust radiation would occur in the far infrared, totally blocked by the Earth's atmosphere and unobservable from ground-based observatories.

Although infrared excess due to dust was detected in NGC 7027 in 1967, the prevalence of dust in planetary nebulae was not widely appreciated until the launch of the *Infrared Astronomical Satellite (IRAS)* in January 1983 (Fig. 10.13). The *IRAS* satellite was a joint mission between the USA, the United Kingdom, and the Netherlands. It carried a 57-cm telescope and was capable of detecting infrared radiation at wavelengths as long as 100 μm. *IRAS* showed that practically all planetary nebulae give off infrared radiation, suggesting that dust is commonly present in planetary nebulae. This view was further strengthened by the observations from the *Infrared Space Observatory* (*ISO*). *ISO* is a *European Space Agency* mission which was launched in November 1995 (Fig. 10.14). Before its coolant ran out in May 1998, *ISO* made detailed observations of dust emission from many planetary nebulae (Fig. 10.15). The dust in planetary nebulae was found to have temperatures of about −150 °C.

The interacting winds theory predicts that the remnants of the asymptotic giant branch wind are still present outside the ionized nebular shell. A schematic diagram of the model is shown in Fig. 10.1. Since the asymptotic giant branch wind contains not only dust but also molecules, would these molecules be also detected in planetary nebulae? The presence of molecules first seemed unlikely

Figure 10.13.
The *IRAS* satellite being launched by a Delta 3910 rocket from Vandenberg Air Force Base on January 25, 1983.

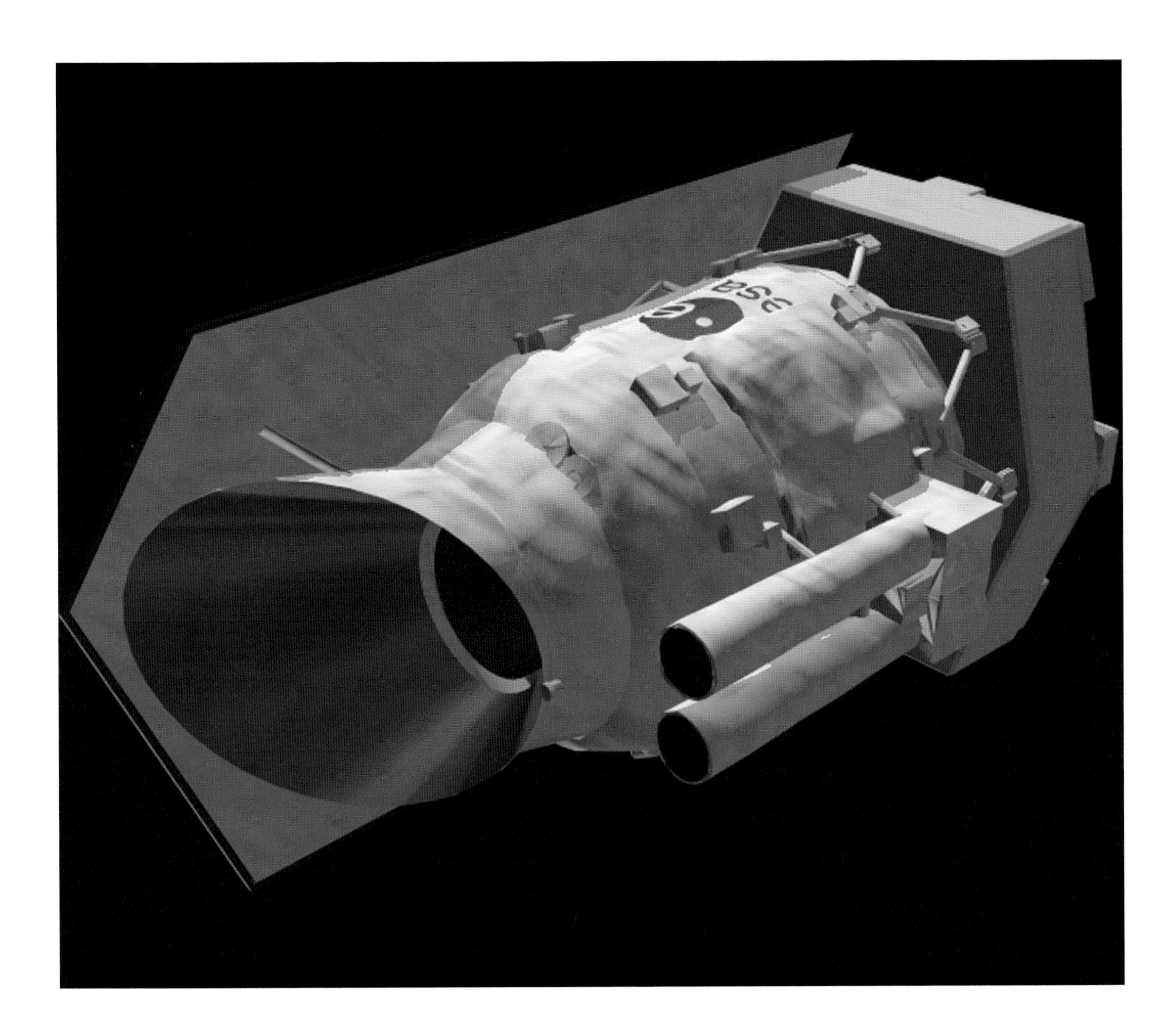
esa

Figure 10.14.
[opposite] The *Infrared Space Observatory* (*ISO*) is a European Space Agency mission which was launched in November 1995. Over its two and half year lifetime, it detected infrared radiation from many celestial sources, including planetary nebulae (credit: European Space Agency).

because the ultraviolet radiation from the hot central stars of planetary nebulae would destroy them. However, molecules including CO and HCN were detected in NGC 7027 and many other planetary nebulae. The shape of these CO lines look exactly the same as those detected in asymptotic giant branch stars such as CW Leo, confirming the evolutionary link between these objects. Not only are there molecules, there are a lot of them. It is estimated that the amount of molecular mass in the halo of NGC 7027 is about 3 solar masses, or about one hundred times the mass contained in the ionized nebular shell!

Another prediction of the interacting winds theory is the existence of X-ray emission from planetary nebulae. When an object is moving faster than the speed of sound, shock waves are generated. In the case of a supersonic aircraft, a cone-shaped shock wave forms at some distance in front of the nose of the plane. The

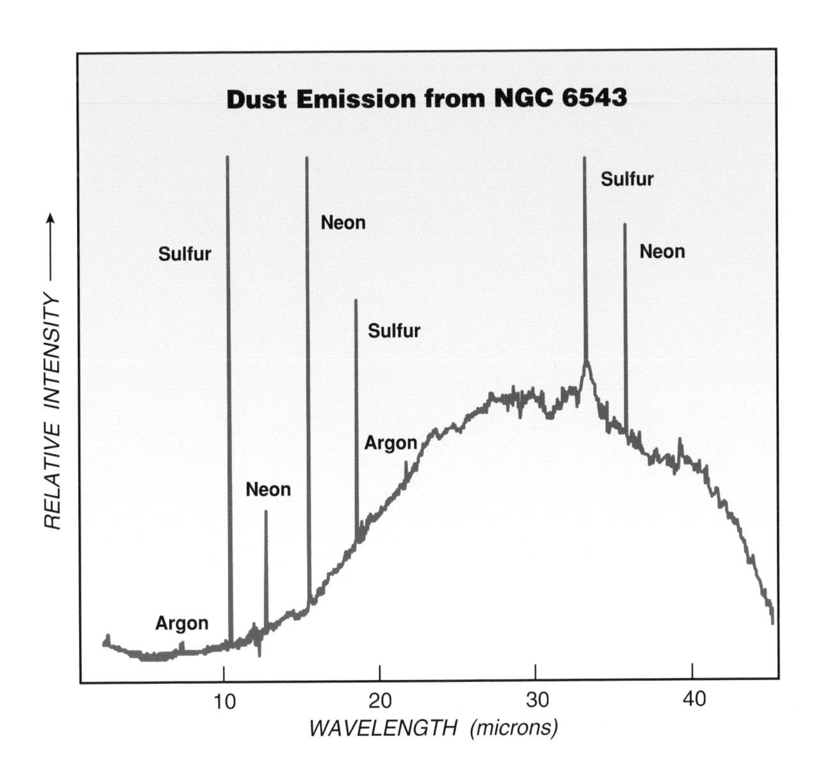

Figure 10.15.
The infrared spectrum of NGC 6543 taken by the *ISO* satellite short wavelength spectrometer. The vertical lines are due to emission lines from atoms and ions. The big hum is due to continuous infrared radiation generated by dust.

fast winds from the central star with velocities of thousands of kilometers per second are highly supersonic (or in aerospace jargon, have a high Mach number). After the passage of a shock wave, the matter left behind is heated up. This creates a bubble inside planetary nebulae with a temperature of millions of degrees. At such high temperatures, X-rays are emitted. This prediction was investigated by the *Roentgen satellite* (*ROSAT*), which made several detections (Fig. 10.16). But due to *ROSAT*'s limited angular resolution, it is not certain that these X-rays originated from the bubbles or from the hot central stars. The definitive confirmation of the existence of the 'hot bubble' had to wait for a new generation of X-ray telescopes.

In 1999, NASA launched the *Advanced X-ray Astrophysical Facility*, the third member of NASA's great space observatories (Figs. 10.17 and 10.18). Re-named *Chandra* in recognition of Chandrasekhar's contributions to high-energy astronomy, it soon demonstrated its superiority over previous X-ray telescopes. At a press conference at the June 2000 meeting of the American Astronomical Society in Rochester, New York, Joel Kastner announced the first *Chandra* image of planetary nebulae. Diffused extended emission was clearly detected in the young planetary nebula BD+30°3639. A month later, at the conference on interacting winds held on the beautiful island of Isle de la Madeleine off the east coast of Canada, You-Hua Chu amazed the participants with *Chandra*'s magnificent X-ray image of the Cat's Eye Nebula (Fig. 10.19). At last, all predictions of the interacting winds had been confirmed by observations.

The mystery of planetary nebulae ejection was finally solved. There is no separate ejection as such, and the nebular shell is just re-arranged wind material. For a hundred years, our attention has been focused on the nebular shell because its higher density makes it brighter in the visible. We now know that this shell represents only a small fraction of the total amount of circumstellar material

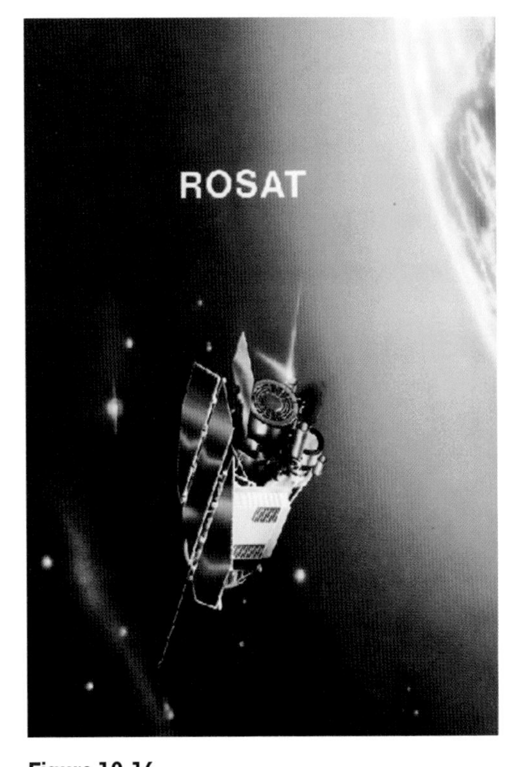

Figure 10.16.
ROSAT is an X-ray space observatory jointly developed by Germany, the US, and the UK. It was launched with a Delta II rocket from Cape Canaveral on June 1, 1990 (credit: Max-Planck-Institut für Extraterrestrische Physik).

Figure 10.17.
[opposite] An artist's illustration of the *Chandra* X-ray Observatory satellite in orbit (TRW). At 15 m long, *Chandra* is the largest satellite that the shuttle has ever launched. The electric power required to operate the *Chandra* spacecraft and instruments is 2 kW, about the same power as a hair dryer (credit: TRW).

USA

Figure 10.18.
[opposite] A picture of the *Chandra* X-ray Observatory (with its Inertial Upper Stage) taken by the space shuttle *Columbia* crew after its deployment on July 23, 1999, 7:47 a.m. This mission, STS-93, was the first NASA shuttle mission commanded by a woman (Col. Eileen Collins). *Chandra*'s unusual orbit was achieved after deployment by a built-in propulsion system which boosted the observatory to a high Earth orbit. This orbit, which has the shape of an ellipse, takes the spacecraft more than a third of the way to the Moon before returning to its closest approach to the Earth of 10 000 km (6200 miles). The time to complete an orbit is 64 h and 18 min. The spacecraft spends 85% of its orbit above the belts of charged particles that surround the Earth. Uninterrupted observations as long as 55 h are possible and the overall percentage of useful observing time is much greater than for the low Earth orbit of a few hundred kilometers used by most satellites (credit: NASA).

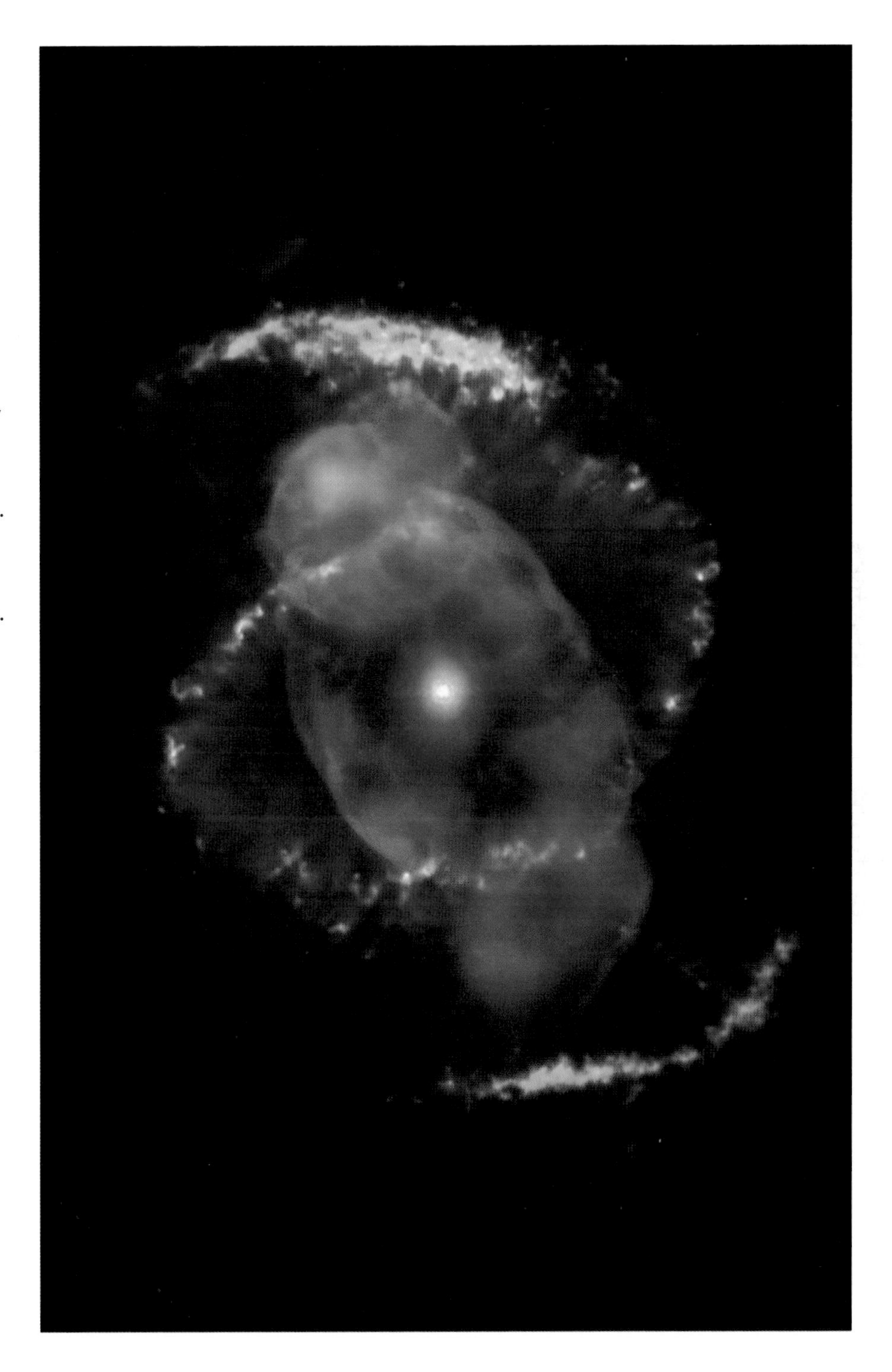

Figure 10.19.
A *Chandra* X-ray image of NGC 6543 superimposed on an *HST* image of the same nebula. The X-ray emitting region is shown in blue. It provides the best illustration of the million-degree 'hot bubble' predicted by the interacting winds model (credit: You-Hua Chu).

in a planetary nebula. The halo, although much fainter, in fact contains over ten or more times the mass of the shell. This discovery also solves the problem of missing mass. There is no doubt that planetary nebulae like NGC 7027 (see Fig. 9.2) must have descended from a star of at least 5 solar masses, therefore giving a definitive proof that stars born with masses much higher than the Chandrasekhar limit will not go supernova. Planetary nebulae, not supernovae, are the common fate of most stars.

11 A morphological menagerie

One of the most puzzling aspects of planetary nebulae is the diversity of their shapes (Figs. 11.1 and 11.2). In 1918, Heber Doust Curtis published a comprehensive catalog of planetary nebulae based on the observations he took with the 36-inch Crossley reflector at Mt. Hamilton. From the 78 drawings and photographs in the catalogue, Curtis classified them into various morphological groups, using descriptive terms such as helical, annular, disk,

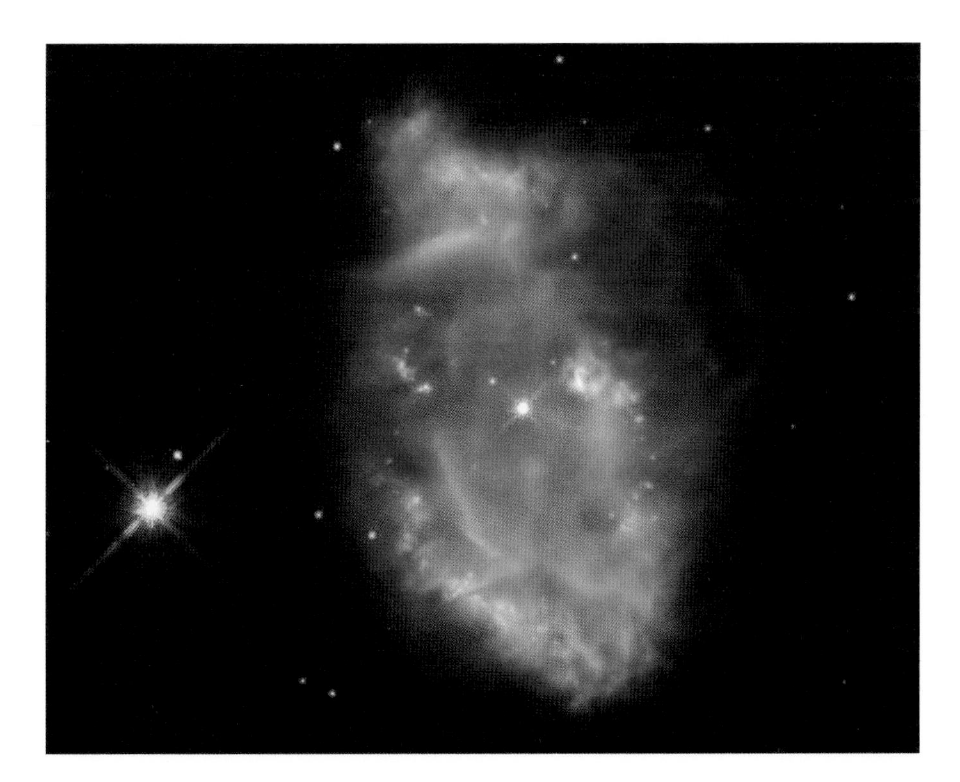

Figure 11.1.
The inner ring of NGC 6309 is surrounded by an outer amorphous structure.

Figure 11.2.
[opposite] NGC 7026 was described by Edward Holden as resembling two sheaves of wheat placed side by side, and was regarded by Lawrence Aller as one of the best examples of the binuclear planetary nebulae. In this *HST* WFPC2 image, we can see a central ring structure with bipolar lobes on each side.

amorphous, and stellar. Since this first attempt, there have been many efforts to classify the morphologies of planetary nebulae. Perek and Kohoutek, in their 1967 catalogue, classified planetary nebulae as stellar, disk, irregular, ring, and anomalous. William Greig of the University of Illinois in 1971 separated planetary nebulae into 15 morphological groups, whereas Ben Zuckerman and Lawrence Aller of UCLA used 16 different classes in their 1986 paper. From a sample of southern planetary nebulae observed at the *European Southern Observatory*, Hugo Schwarz, Romano Corradi and Letizia Stanghellini reduced the number of classes to five: elliptical, bipolar, point-symmetric, irregular, and stellar. Arturo Manchado of the Instituto de Astrofisica de Canarias reduced the number further to three: round, elliptical, and bipolar.

All these morphological classifications seem frustratingly confusing. Is there an order among this chaos? While we can imagine that elliptical nebulae may somehow be related to round nebulae, there is a class of planetary nebulae that is totally different in appearance: the butterflies. Figures 11.3–11.9 show some of the

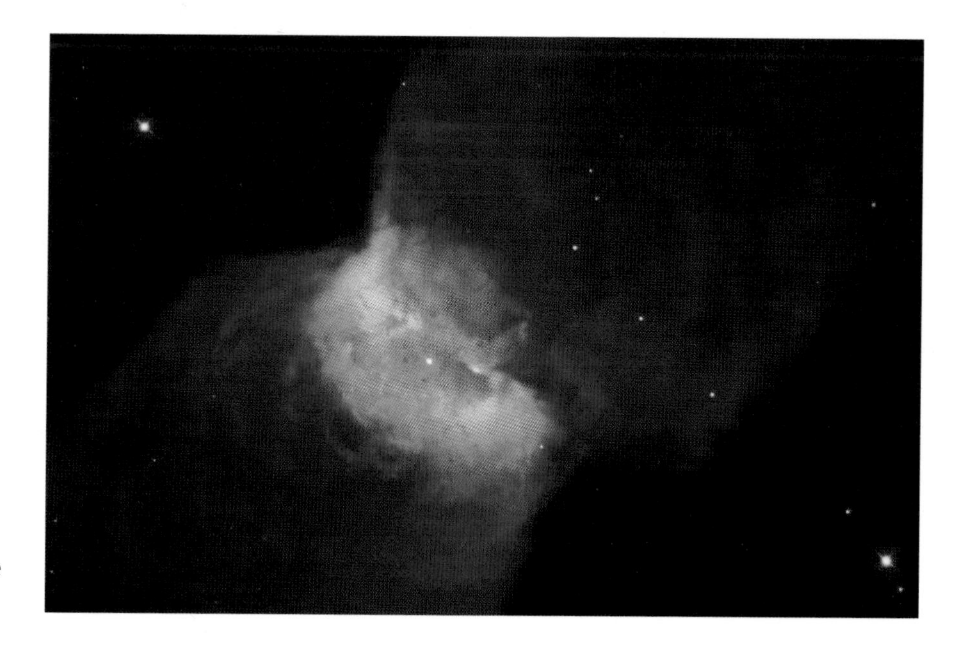

Figure 11.3.
The lobes of the bipolar planetary nebula NGC 2346 in Monoceros extend over almost 3 arcmin in the sky (credit: NASA and the Hubble Heritage Team, Space Telescope Science Institute).

best examples of the butterfly class. This class of objects is defined by the existence of a pair of lobes with a perfect axis of symmetry. Like butterfly wings, the lobes are wide at the ends with a narrow center (see also Fig. 11.10).

Figure 11.4.
NGC 6302 in Scorpius is probably the most beautiful butterfly in the sky. The two lobes are separated by a dust torus, which obscures the central star. The *IRAS* and *ISO* spacecrafts have detected emission lines from neon atoms with five electrons removed and silicon atoms with five and six electrons removed. In order to ionize these atoms to such a high degree, very energetic ultraviolet photons from the central star are required (see Chapter 3). This implies a very hot central star. From the *IRAS* and *ISO* data, Stuart Pottasch estimated the central star temperature to be 380 000 °C, making it one of the hottest stars known in the Universe. This picture was taken by Trung Hua with the *Siding Spring Observatory* 2.3-m telescope (credit: Hua).

Figure 11.5.
Hubble 5 in Sagittarius is one of the most beautiful bipolar planetary nebulae, with lobes extending to half an arcminute on each side. This *HST* image reveals a complex network of loops and filaments in the core and the lobes. The central star is not visible, probably hidden behind the dust torus separating the bipolar lobes.

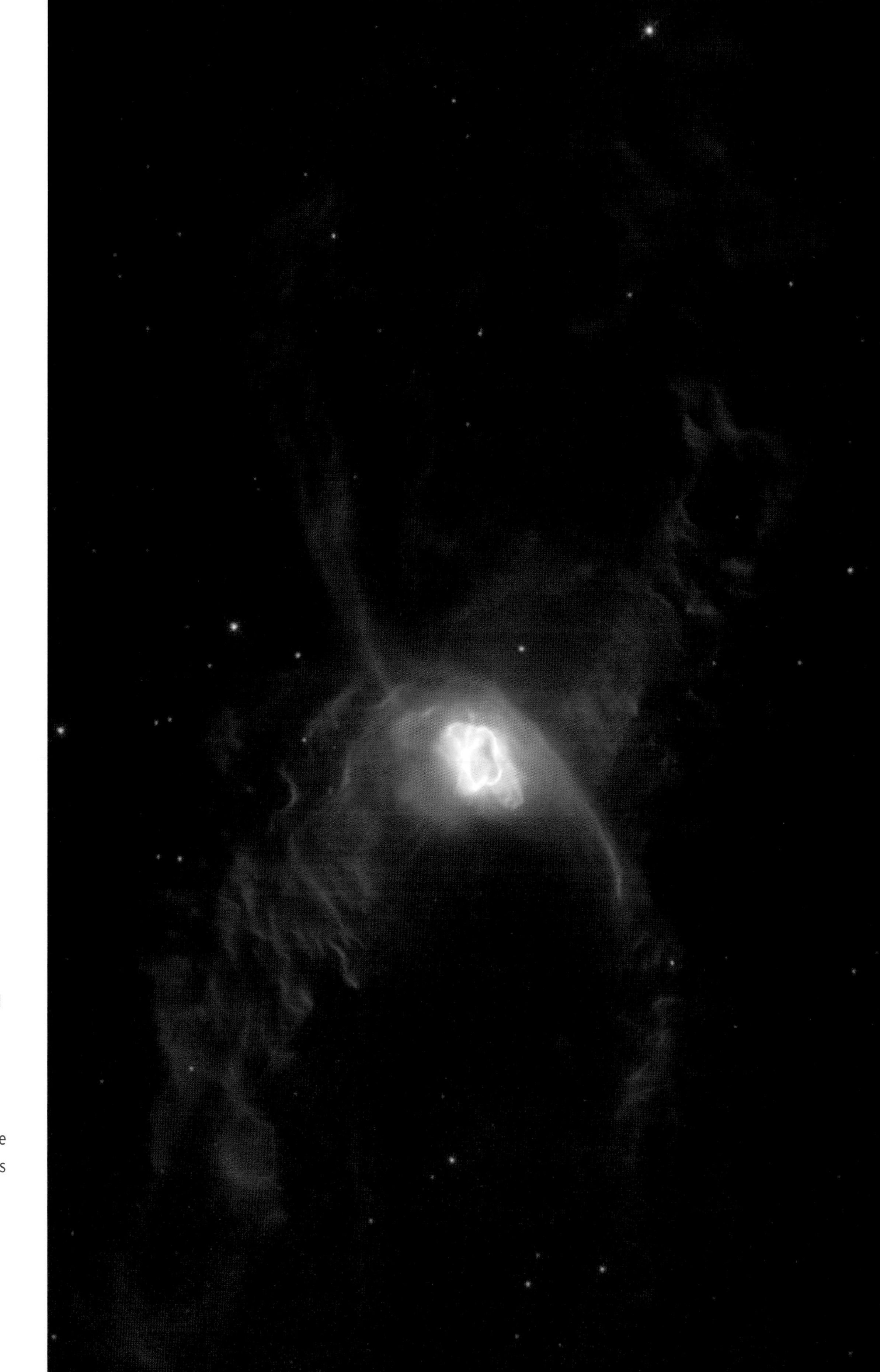

Figure 11.6.
Although NGC 6537 was discovered by Edward Charles Pickering as early as 1882, the bipolar lobes outside the bright central core were not known until 1990. They were found by Angels Riera as part of his PhD thesis at Instituto de Astrofisica de Canarias. In this *HST* picture, the two lobes of NGC 6537 can be seen extending as far as 2 arcmin on each side, looking like the claws of a crab.

We have to remember that planetary nebulae are defined by their spectrum. So in spite of the very different appearances of butterfly nebulae, there is no doubt that they are planetary nebulae. If this is the case, how do we account for their unique shapes? What about those that are not exactly round, but are elliptical to various degrees? Is there an intrinsic structure that is inherent in all these seemingly diverse morphologies?

This question may be best answered using the most famous object that characterizes everyone's definition of a planetary nebula: the Ring Nebula (M57, NGC 6720) in Lyra. The Ring Nebula was one of the first planetary nebulae to be discovered, and is also one of the four planetary nebulae included in Charles Messier's 1784 catalog of nebulous objects (see Chapter 1). Because of its well-defined ring-like appearance, the Ring Nebula is often held up as the classical example of a planetary nebula. As planetary nebulae go, the Ring Nebula is relatively nearby; its distance is estimated to be between 1000 and 2000 light years. At that distance, the size of the nebula is approximately 500 times the size of our solar system.

The quest to decipher the structure of the Ring Nebula began with Curtis. For the Ring Nebula and other objects with annular (ring-like) structures, the most obvious model is that the ring represents the projection of a hollow spherical shell on the sky. The edge of the shell is brighter because we look through more material along the edges compared to a line of sight straight through the middle. However, a quantitative evaluation suggests otherwise. The brightness contrast between the ring and the hollow center is too large. If it were a spherical shell, the front and back of the shell would project onto the center, and the center would be brighter than it is. For a spherical shell, the central hole should be approximately half as bright as the ring, but instead Curtis found a brightness ratio of about 1:20. Curtis therefore came to the conclusion that the Ring Nebula is not a spherical shell, but a true two-dimensional ring lying flat on the plane of the sky.

Figure 11.7.
NGC 6881 bears a strong resemblance to NGC 6537, except that it is much smaller (7 arcsec).

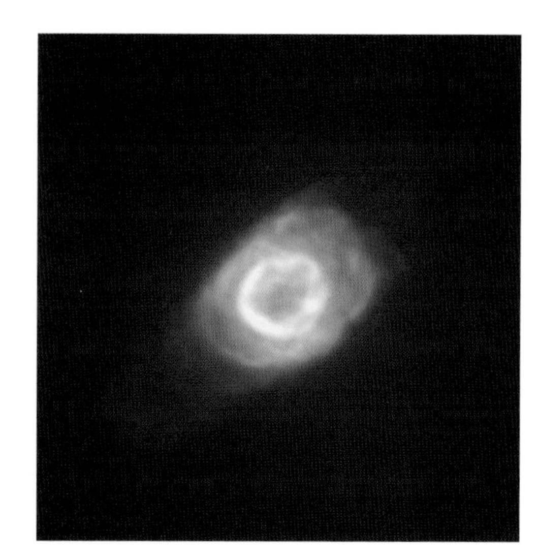

Figure 11.8.
NGC 6790 has long been known from ground-based optical and radio observations to have a ring-like structure. Its outer butterfly wings are only revealed when observed by the *HST*.

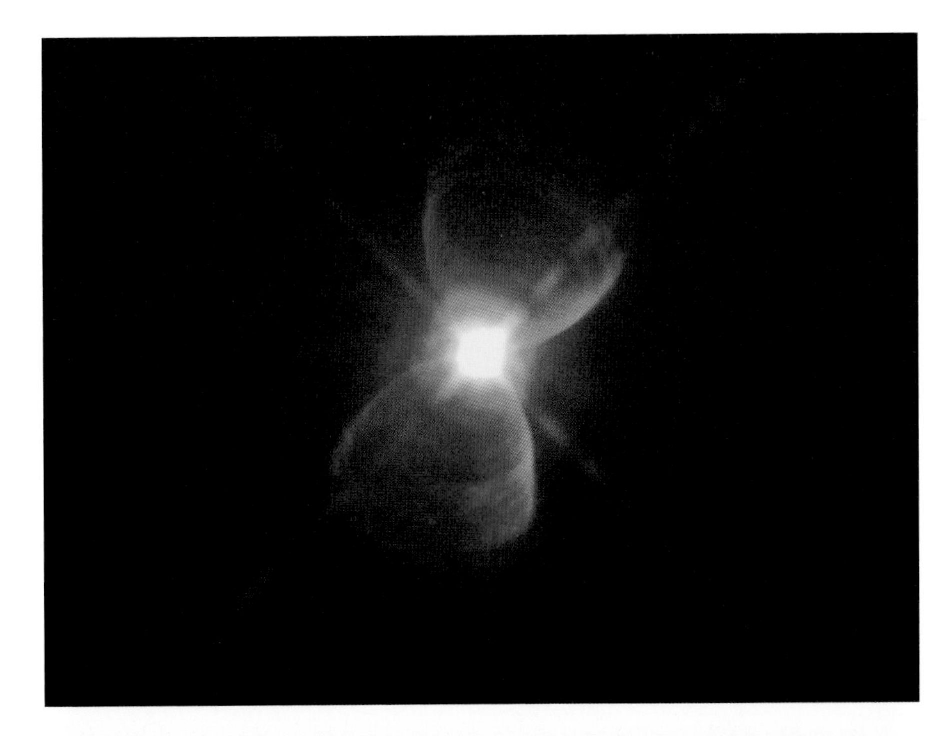

Figure 11.9.
Hb12 is a small, young planetary nebula similar in structure to NGC 2346. For its young age and bipolar morphology, Joseph Hora and Bill Latter call it 'butterfly in the making'.

Figure 11.10.
He2-104 has been referred to as the 'Southern Crab' for its appearance. It was discovered by Karl Henize as an emission-line object in his southern sky survey and later confirmed as a planetary nebula by Louise Webster of *Mt. Stromlo Observatory*. It has a clear bipolar structure in this *HST* WFPC2 image (credit: R. Corradi).

If this is the case, and if the Ring Nebula is not unique among planetary nebulae, then there should exist other planetary nebulae of similar ring shape but different orientation. For example, if the ring were aligned perpendicular to the plane of the sky, its appearance would be very different. Curtis suggested that NGC 650-1 (M76), which has a roughly rectangular shape, could in fact represent such a case. It has the appearance of a nebulous bar of about $1^1/_2$ by $^3/_4$ arcmin in size (Fig. 11.11). Just imagine holding a torus containing colored liquid edge-on against a light. The two ends would appear darker because light would have to pass through

Figure 11.11.
This two-minute exposure of NGC 650-1 was taken at the *WIYN* telescope at *Kitt Peak*, Arizona. The object was given two NGC numbers because William Herschel thought it was an unresolved double nebula (credit: A. McDonald/WIYN/National Optical Astronomy Observatory/National Science Foundation).

more material. The two ends of NGC 650-1 appear brighter exactly for the same reason, except that we are dealing with a gaseous torus that emits light rather than a liquid one that absorbs light.

This problem was picked up 40 years later by Rudolph Minkowski. Minkowski and Don Osterbrock photographed the Ring Nebula and NGC 650-1 with the 200-inch telescope of *Mt. Wilson* and *Palomar Observatories*. They concluded that the two objects have the same intrinsic shape of a flattened ring, more or less like a donut. In the case of the Ring Nebula, the donut lies approximately 45° from the plane of the sky, but for NGC 650-1, the donut is seen edge on against the plane of the sky. This was viewed as a great achievement at the time, until further observations unveiled some complications.

In the 1970s, Raymond Louise and Trung Hua of the *Observatorie de Marseille* used the 193-cm telescope at *Haute Provence Observatory* in southern France to take images of the Ring Nebula. By placing filters with narrow bandpasses in front of the camera, they allowed light from only one particular atom to pass through. They found that the nebula looks quite different when observed with different filters. They proposed that the Ring Nebula is an oblate spheroid with a higher density in the equatorial regions: in other words, a compromise between the uniform spheroid and the donut models.

This approach was taken a step further by N.K. Reay and Sue Worswick of the Imperial College, who in 1975 observed the Ring Nebula with more filters using an electronograph camera mounted on the 42-inch telescope at *Lowell Observatory* in Arizona. They found that while in the light of neutral oxygen and singly ionized nitrogen, the ring is brighter than the center by a factor of 30; however, this ratio becomes only 2:1 in the light of singly ionized helium (Fig. 11.12). This means that the nebula takes on different shapes under different light! It resembles a spheroid if we look at only the helium emission and a donut when observed in other lines.

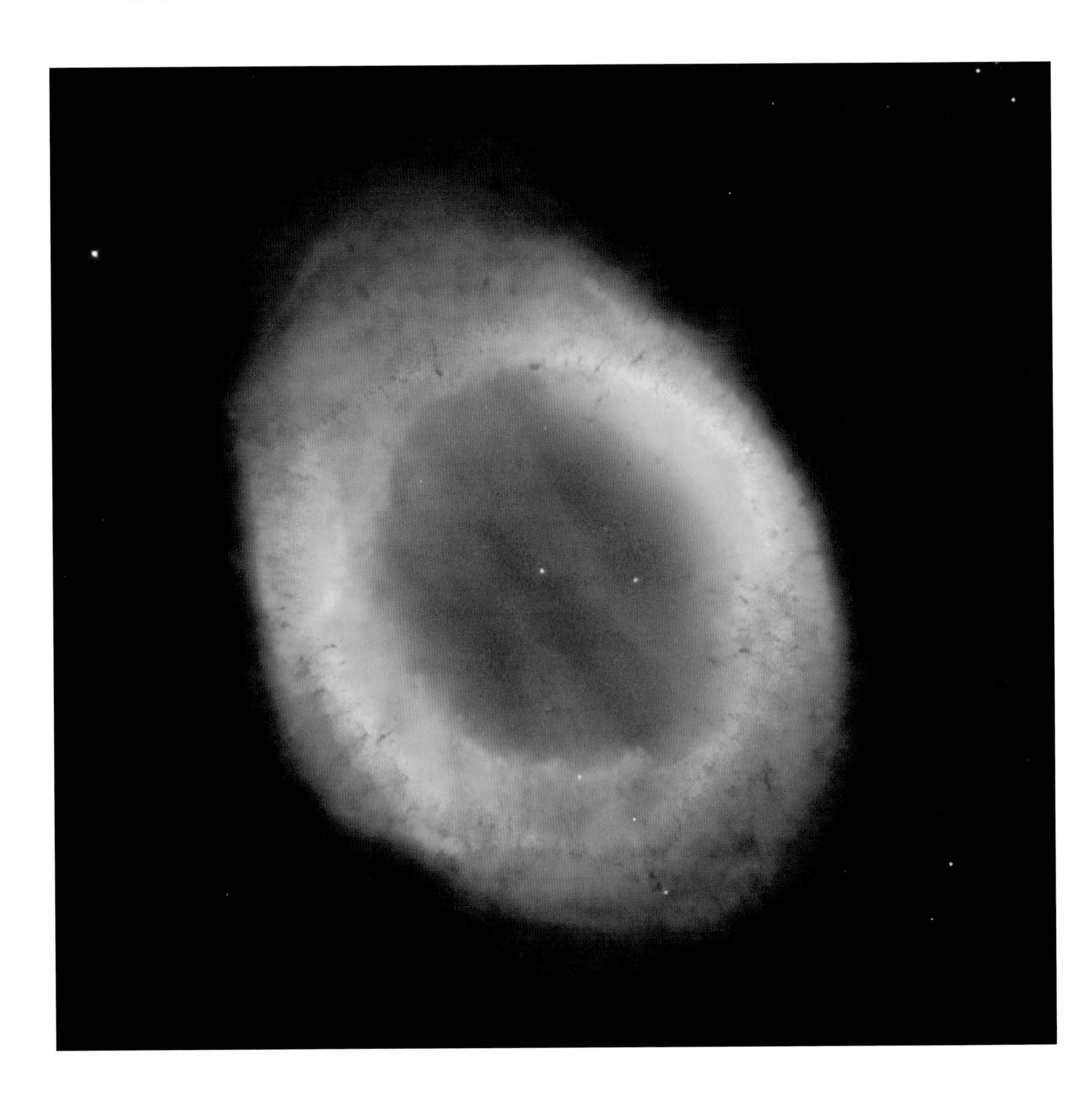

Figure 11.12.
[opposite] *HST* Wide Field Planetary Camera 2 image of the Ring Nebula. We can see the light from the helium atom (shown in blue) is primarily confined to the center of the ring, whereas the extent of singly ionized nitrogen (in red) is much larger. Green color corresponds to doubly ionized oxygen (credit: NASA and the Hubble Heritage Team, Space Telescope Science Institute).

This is because different lines are excited under different physical conditions. Helium atoms, requiring higher energy photons to ionize than hydrogen or oxygen, confine their emissions close to the star (Fig. 11.13). The shape of a planetary nebula is therefore not a geometric concept like a teacup or saucer (which looks the same regardless of color) but is dependent on physical processes occurring in the nebula.

The debate became heated in the mid-1970s as more players entered into the ring. P.E. Proisy of the *Observatoire de Lyon* suggested that the Ring Nebula is a cylinder, with its major axis lying at 30–50 ° from our line of sight. L.E. Goad, in his Ph.D. thesis entitled 'The Structure of Planetary Nebulae' from Harvard University in 1975, came to similar conclusions. Peter Kupferman, in his PhD thesis at the University of Michigan in 1981, returned to the spheroid model with a twist. In his model, while the geometry remains a spheroid, the density inside is

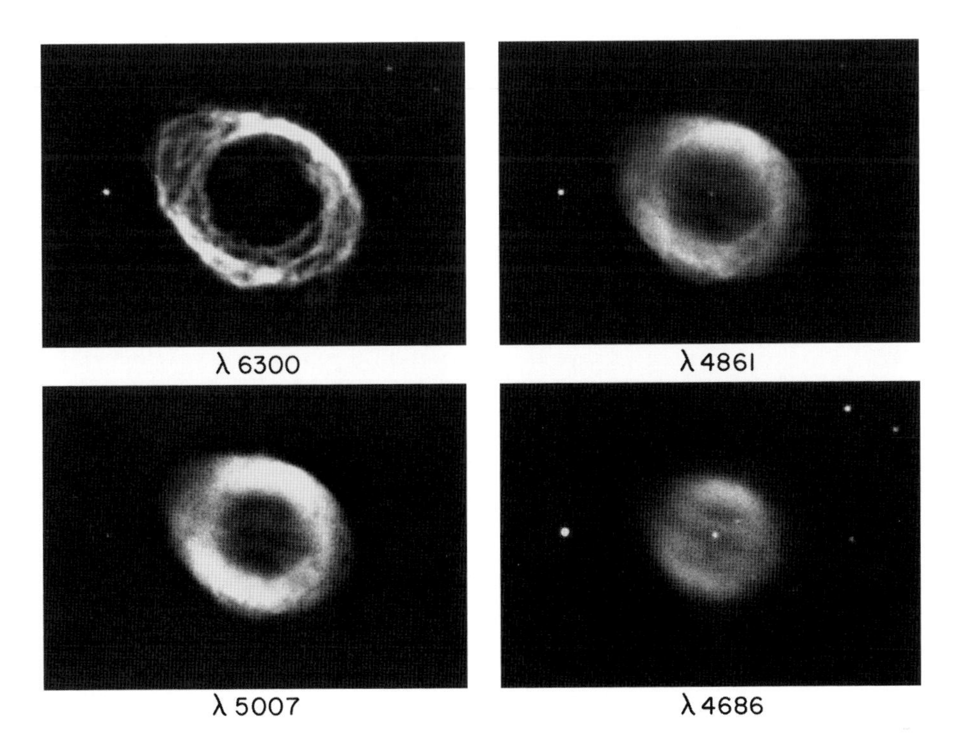

Figure 11.13.
The images of the Ring Nebula in four different filters: neutral oxygen (upper left), neutral hydrogen (upper right), doubly ionized oxygen (lower left), singly ionized helium (lower right) (credit: National Optical Astronomical Observatory/National Science Foundation).

variable, reaching a minimum along the line of sight. Both the cylindrical model and the variable-density spheroid model were able to quantitatively match the brightness variation across the nebula.

These geometric models are successful when suitable artificial parameters are provided, but they do not answer the fundamental question: why does the Ring Nebula take on this particular shape? If this shape is produced by a physical process, then what is this process? Is the Ring Nebula unique or is its geometry typical of other planetary nebulae?

Although the elliptical ring is the dominant structure of the Ring Nebula, it also has peripheral structures that were not taken into consideration in the models. On August 6, 1936, John C. Duncan photographed the Ring Nebula with the 100-inch telescope at *Mt. Wilson Observatory*. To his surprise, he found a previously unseen faint envelope outside the ring. A modern CCD picture of the Ring Nebula is shown in Fig. 11.14. The outer envelope, with a semi-major radius of 78 arcsec, seems to consist of loops and filaments protruding from the inner shell. Further out, there is an even fainter circular halo of radius 115 arcsec. What is the origin of these halos, and how are they related to the inner ring?

Figure 11.14.
A deep image of the Ring Nebula taken by George Jacoby at *Kitt Peak National Observatory* showing the faint outer halos first discovered by Duncan in 1936. The inner parts of the nebula are overposed to reveal the faint halo (credit: National Optical Astronomy Observatory/ National Science Foundation).

As we saw in the previous chapter, outer halos in planetary nebulae are not uncommon at all. The detection of halos in planetary nebulae gave a boost to the interacting winds theory. A later extension of this theory by Bruce Balick of the University of Washington suggests that if the slow wind is not spherically symmetric, the interacting winds process can in fact create different morphologies of planetary nebulae as observed.

The interacting winds theory was the first physical theory that successfully explained the formation and expansion of planetary nebulae. However, lingering questions remained. For example, the Ring Nebula has two halos. Do both halos share the same origin or are they manifestations of different physical processes?

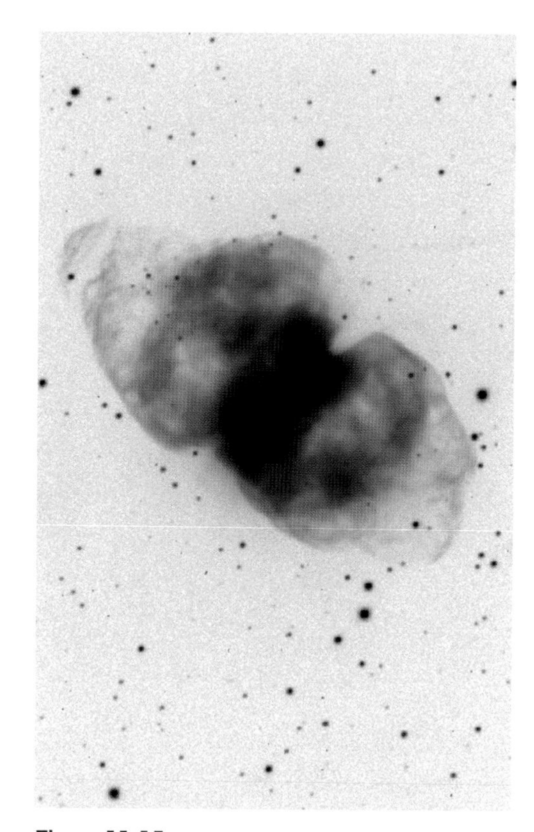

Figure 11.15.
A CCD image of NGC 650-1 taken by George Jacoby at *Kitt Peak National Observatory*. The faint outer lobes are difficult to see in photographic plates (credit: National Optical Astronomy Observatory/National Science Foundation).

CCD observations also revealed other faint structures not previously seen in images taken with photographic plates. Remember NGC 650-1, the nebula that Curtis and Minkowski thought was the Ring Nebula turned sideways? In Fig. 11.15 we show a CCD image of NGC 650-1. In addition to the rectangular shape that Curtis saw, it consists of two faint outer lobes extending over almost 5 arcmin in total extent. Most remarkably, the major axes of the bright and faint structures are off exactly by 90°! Bruce Balick suggested that there are unseen high-density, neutral materials outside of the bright core. This dark matter confines the central donut and forces any outflow from the star to emerge through the donut hole, creating the bipolar lobes observed.

The existence of planetary nebulae with bipolar shapes has been known for a long time. NGC 2346 (Fig. 11.3) and NGC 6302 (Fig. 11.4) are good examples. However, bipolar nebulae were thought to be rare, making up no more than 10% of the total population of planetary nebulae (the majority being of ring or elliptical shape). The case of NGC 650-1, however, gave us reasons to be concerned. How many bipolar planetary nebulae had we missed because we had not gone deep enough?

Astronomers were surprised to find that other planetary nebulae also take on a very different look when a deeper picture is taken. One example is Sh1-89 (Fig. 11.16). This planetary nebula was first discovered as a nebulous object on the plates obtained with the 48-inch Schmidt telescope in *Mt. Wilson*. Stewart Sharpless, a Carnegie Fellow at *Mt. Wilson Observatory*, classified its morphology as 'irregular'. When imaged on photographic plates, only the central bright core can be seen, which has a rectangular appearance. However, a deep CCD image reveals the fainter lobes on both sides of the rectangle. The nebula is bipolar!

So maybe bipolar planetary nebulae are more common than we thought (Fig. 11.17). If that is the case, what would a bipolar nebula look like if turned sideways? From the images of Sh 1-89 and

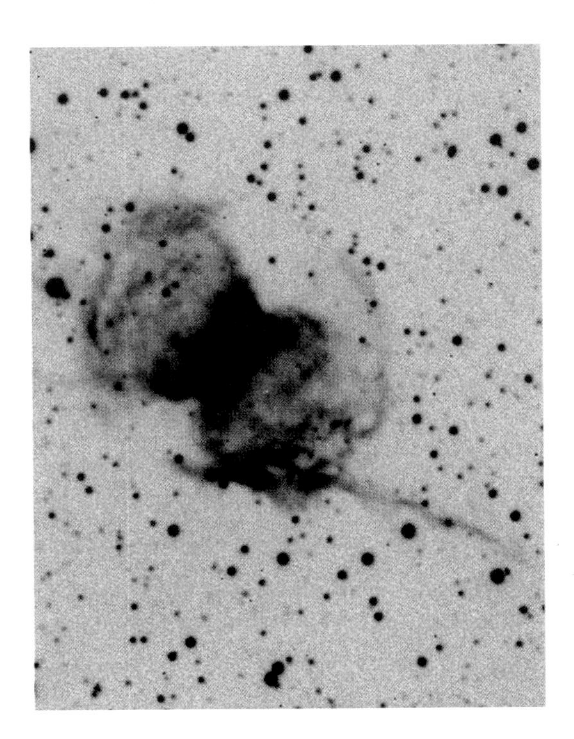

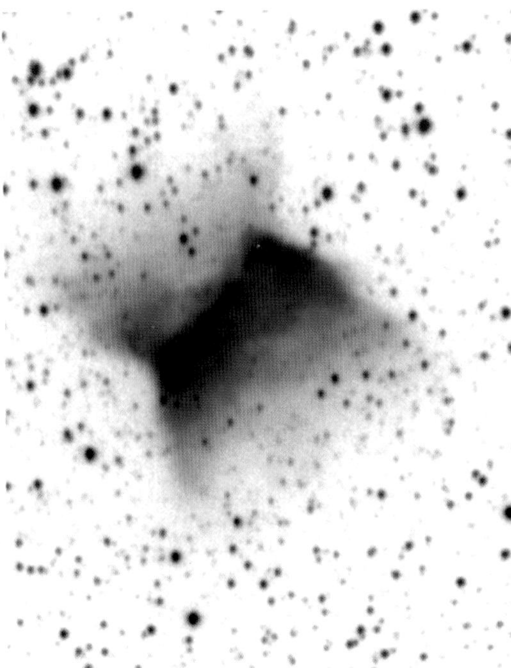

Figure 11.16.
[left] An image of Sh1-89 taken in the light of singly ionized nitrogen by Trung Hua with the 1.2-m telescope of the *Haute Provence Observatory*. The central bright part is a donut seen edge on (credit: Trung Hua).

Figure 11.17.
[right] SaWe3 is another example of a planetary nebula found to be bipolar only after a deep exposure is taken. When only the bright core was visible, the nebula was thought to have a rectangular shape (credit: Trung Hua).

NGC 2346, we can see that the ends of the lobes are larger than their bases. If we observe them with a line of sight down the major axis of the lobes, the front and back lobes are projected onto the plane of the sky as filamentary circular blobs, not unlike the halos that we see in the Ring Nebula! It is therefore not inconceivable that the Ring Nebula is in fact a bipolar nebula (Fig. 11.18).

Although this view is still controversial, I believe that most, if not all, planetary nebulae with ring shape are intrinsically bipolar. Through computer models, H. Monteiro of the Instituto Astronômico e Geofisico of Brazil and Christophe Morisset of the Laboratorie d'Astronomie Spatiale of France have found that the famous Southern Ring NGC 3132 (Fig. 1.9) has a diabolo shape in three dimensions. Owing to the lack of spatial resolution and brightness sensitivity, we have been distracted by the bright torus (namely the ring) and did not appreciate the faint bipolar lobes that could be present in all ring-like or even elliptical planetary nebulae.

Figure 11.18.
[opposite] A false color Hα image (25 min exposure) of the Ring Nebula taken with the 8.3-m *Subaru Telescope* on Mauna Kea. Two different halos can be seen. The inner oval-shape halo (size 160 $\times$ 145 arcsec) is due to projection of the front and back lobes whereas the outer circular halo (size 230 arcsec) is the remnant of the slow red giant wind (credit: copyright © 2000 Subaru Telescope, National Astronomical Observatory of Japan, all rights reserved).

Do we finally have the truth about the nature of the Ring Nebula? It is difficult to say, but surely we have learned some very valuable lessons. The goal of science is to reduce the apparent chaos and complexity of nature to simple basic principles. We have accumulated a lot of evidence suggesting that rings and butterflies are the same thing. It would be truly amazing if all the animals in the planetary nebulae zoo could be explained by this simple model.

12 Butterflies in the sky

In the previous chapter, we found that bipolar nebulae are probably more common than astronomers first believed. If this is the case, it would be useful for us to study the structures of bipolar nebulae in detail in the hope of getting a deeper understanding of the structure of planetary nebulae. What is particularly striking about the pictures of bipolar nebulae such as NGC 2346 (Fig. 11.3) and NGC 6302 (Fig. 11.4) is the 'tight waist' separating the two lobes. The pictures give the impression that the lobes are breaking out of a confined space. The confinement is provided by a donut-shaped torus, which forces the lobes to go out from the open ends.

Maybe the different morphologies that we see in planetary nebulae are an illusion. They all may have the same intrinsic structure after all. Astronomers have long been puzzled by the variety of shapes seen in planetary nebulae. It is tempting to believe that such diverse morphologies are just different manifestations of a single, unified, basic three-dimensional structure. The first attempt to characterize such a structure was made by the Russian astronomer Gabriel Khromov and the Czech astronomer Lubos Kohoutek. They emphasized the effect of projection on the apparent structure of planetary nebulae and suggested in 1968 that many of the observed morphologies can be explained by cylinders with open ends projected at different orientations on the sky (Fig. 12.1).

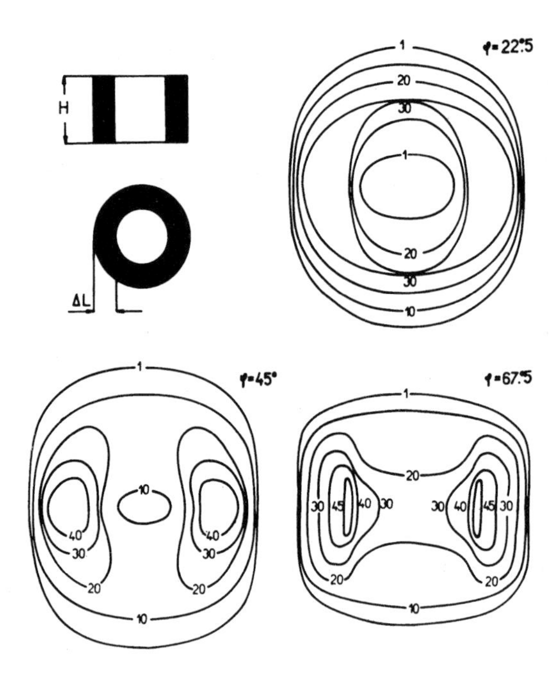

Figure 12.1.
The open-cylinder model of Khromov and Kohoutek. When gaseous cylinders of diameter *L*, height *H*, and thickness ΔL (side and top view shown on the left) are viewed at different inclination angles (22.5°, 45°, and 67.5° for the examples on the right), they resemble the observed morphologies of planetary nebulae (credit: G.S. Khromov and L. Kohoutek, *Bulletin of the Astronomical Institutes of Czechoslovakia*, **19**, No. 2).

Another attempt to create a universal model of planetary nebulae was made by Colin Masson of the Center for Astrophysics in 1990. He proposed that the structure of the Ring Nebula is consistent with a spherical shell with both radial and latitude-dependent density gradients. Since the central star has only a finite output of ultraviolet photons, the nebula is ionized to different depths along different directions (Fig. 12.2). When projected to the plane of the sky and viewed at different angles, the ionized shell can have shapes ranging from rings to hourglasses.

Although the Masson model has more parameters than the previous models, its advantage lies in its introduction of physics (photoionization) into the problem. In order to test the power of the Masson model, my graduate student Orla Aaquist fitted many radio images of planetary nebulae with the model. In 1998 Cheng-yue Zhang and I made a more ambitious effort and produced fittings for 110 planetary nebulae. Some examples of the simulated images can be found in Fig. 12.3.

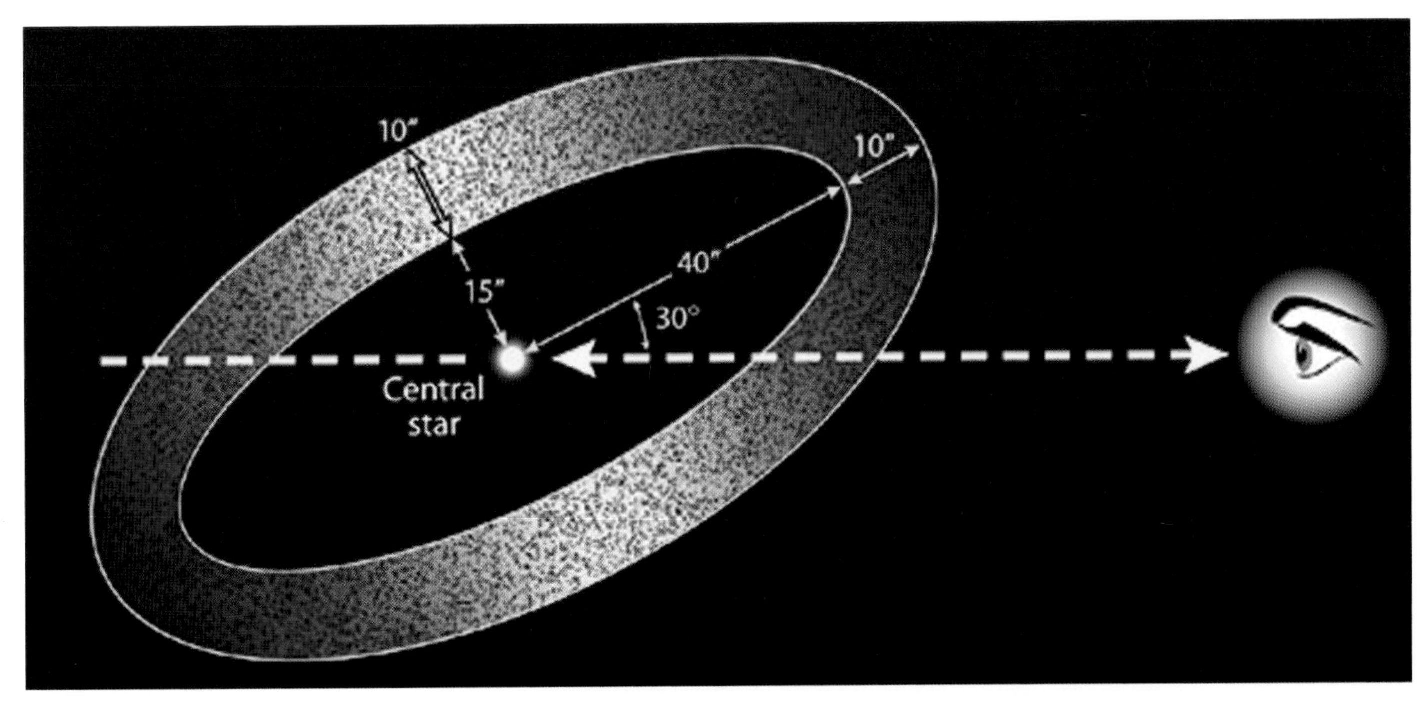

Figure 12.2.
[above] A schematic diagram of the ellipsoidal shell model of Masson. The Ring Nebula is modeled by an ellipsoidal shell of uniform thickness (10 arcsec). The semi-major and semi-minor inner radii are 40 and 15 arcsec respectively. The darker shades correspond to regions of higher density. The lines of sight to the nebula are represented by the horizontal lines (diagram adapted from Masson, C.R. 1990, *Astrophysical Journal*, **348**, 580).

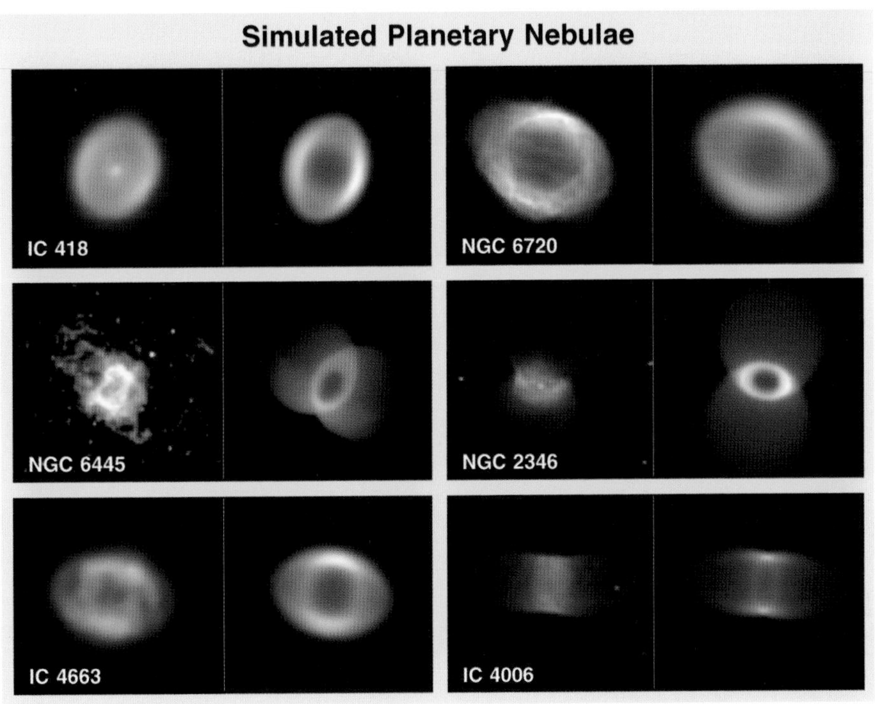

Figure 12.3.
[right] Comparisons between the observed (left) and simulated (right) images of six planetary nebulae. The success of these simulations suggests that planetary nebulae have similar intrinsic structures and that the differences in apparent structures among planetary nebulae are in part due to their orientations in the sky.

Given the fact that Khromov and Kohoutek did not have the benefits of the CCD images available today, their idea was remarkably insightful. We now believe that the brightest part of the nebula has the shape of a short cylinder (a torus or a donut), but the polar extensions have the shape of fans or lobes. The finite extent of the torus is determined by ionization effects. The lobes, however, are created by a combination of dynamical and ionization effects.

A schematic picture of a universal model of planetary nebulae is shown in Fig. 12.4. The torus, having higher densities, forces the

Figure 12.4.
A schematic model of planetary nebulae. The halo (in dark red) represents the remnant of the slow wind from its red giant progenitor. This wind is denser along the equatorial directions, and the torus represents part of this density enhancement. The outer edge of the torus is defined by the ionization limit. Because of lower density in the polar directions, the fast wind can break out more easily, creating the two lobes.

outflow from the star to channel through the open ends. After breaking out, the outflow fans out to the side, giving the planetary nebulae a butterfly shape. Looking down the symmetry axis, the nebula looks like a ring with faint halos, exactly like the Ring Nebula. If we rotate this model to different orientations, then the projection to the plane of the sky will give rise to different morphologies.

If we accept this universal model of planetary nebulae, the next question is: why? How do planetary nebulae come to have this form? If the planetary nebulae originate from the circumstellar envelopes of red giants, which are by-and-large round, how do the torus and lobes develop?

The answer may be found in the interacting winds process. In 1987, Bruce Balick suggested that the asymptotic giant branch star envelopes may have a slight density asymmetry. When the fast wind begins, it will then exploit the weak points in the envelope and push further along those directions. This could create the bipolar lobes that are observed. When this fast outflow runs into the remnant of the asymptotic giant branch wind, shocks are generated. The signature of shocks can be found by an emission line of molecular hydrogen (H_2) at the wavelength of 2.12 μm. Figure 12.5 is an infrared image of the Ring Nebula taken with a molecular hydrogen filter. The streams and filaments that we see in this picture could be evidence of shock interactions.

In Balick's picture, planetary nebulae begin their lives with round and spherical appearances and develop into increasingly asymmetric forms as they age. This theory has inspired two magnificent PhD theses, one by Adam Frank from the University of Washington and another by Garrelt Mellema of the University of Leiden. Frank and Mellema were able to use the interacting winds theory to produce simulated images that are difficult to tell from the real ones. There is no doubt that interacting winds play an important role in the shaping of planetary nebulae (Fig. 12.7).

Figure 12.5.
[opposite] Streams of outflowing gas can be seen in this infrared image of the Ring Nebula taken by David Thompson at *Calar Alto Observatory* in the Sierra de Los Filabres, Spain. The filter is centered around the 2.12 μm emission line of molecular hydrogen.

Figure 12.6.
[right] The red glow in this near-infrared picture of NGC 6781 is due to emission from molecular hydrogen at a wavelength of 2.12 μm. This emission line is generated by shock waves as a result of the interacting winds process. In this picture, the colors blue, green and red are used to represent the near-infrared colors at 1.2, 1.6 and 2.2 μm. This image was obtained as part of the Two Micron All Sky Survey (2MASS) carried out by the University of Massachusetts and the Infrared Processing and Analysis Center of the California Institute of Technology. Two 1.3-m telescopes, one in the northern hemisphere at Mt. Hopkins, Arizona, and another in the southern hemisphere at Cerro Tololo, Chile, were used for the survey.

In spite of these successes, some fundamental questions remain. Balick's theory requires that the red giant envelope have a certain degree of asymmetry which is later amplified by the interacting winds process. What is the cause of such asymmetry?

Another prediction of Balick's theory is that the morphology of planetary nebulae is dependent on age. Young nebulae should be statistically rounder in comparison with the older population. This prediction can be tested by observing very young planetary nebulae. If they already possess bipolar forms, then they must have been born this way and the interacting winds process only enhances the

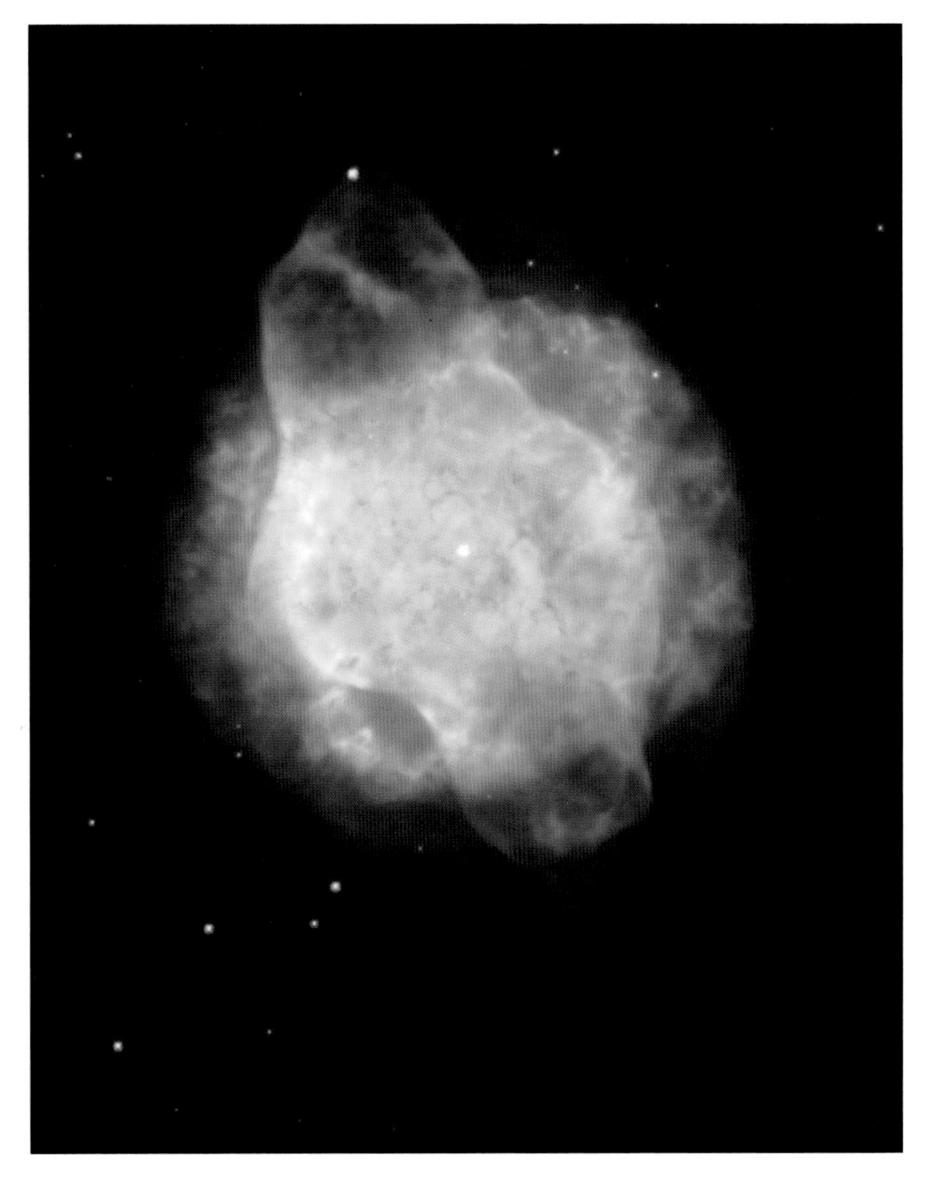

Figure 12.7.
NGC 3918 shows an inner shell of roughly elliptical shape surrounded by an outer halo of 16 arcsec diameter. A pair of lobes in a spindle shape extends beyond the halo, a result of the fast wind breaking through the remnant of the red giant wind. Romando Corradi of the Instituto de Astrofisica de Canarias, who made extensive observations of NGC 3918 with the 3.5-m New Technology Telescope of the *European Southern Observatory*, compared his observational results to the theoretical models by Mellema and found them to be in excellent agreement. NGC 3918 is one of many examples showing that the morphologies of planetary nebulae are shaped by the interacting winds process.

asymmetry instead of creating it. The key to test this theory therefore lies in the observations of the youngest planetary nebulae, or better still, the nebular shape before the birth of planetary nebulae. The search for such objects, called proto-planetary nebulae, is the subject of the next chapter.

13 The missing link

In Chapter 3 we learned that planetary nebulae are lit up by ultraviolet photons emitted by the central star. However, significant amounts of ultraviolet light are emitted only when the central star is hot enough, at least 30 000 °C. So even if the nebular shell has been formed, it is difficult to see. According to the stellar evolutionary model of Schönberner, there is a time gap of about 3000 years between the end of the asymptotic giant branch and the beginning of the planetary nebula phase. This is the time needed for the star to evolve from 3000 °C to 30 000 °C as the hot core is further exposed. The objects in transition, often referred to as proto-planetary nebulae, are a key missing link in our understanding of stellar evolution.

Finding examples of this link at first seemed hopeless. A star at an intermediate stage would hardly look out of the ordinary. And with only a few hundred expected in the Galaxy, proto-planetary nebulae would be impossible to locate among the Milky Way's saturated star fields.

I became interested in the search for these transition objects in the early 1980s. From infrared observations, I knew that asymptotic giant branch stars wrap themselves in tight dust cocoons. Even when the mass loss process stops at the end of the asymptotic giant branch phase, their dust cocoons still remain. In fact, the remnants of such cocoons should still be there during the proto-planetary nebula phase and even several thousand years later in the planetary nebula stage. Since the dust is dispersing and cooling, these remnants could still be detectable in planetary nebulae by far infrared observations. The chance of detecting such remnants would be the best in young planetary nebulae before the dust disperses too much.

Unfortunately, the Earth's atmosphere is completely opaque to far infrared radiation, and the search for such dust cocoons is impossible with ground-based telescopes. The first survey of the mid-infrared sky was made by the *Air Force Geophysical Laboratory* (*AFGL*). Between 1971 and 1974, a total of nine rocket flights were launched: seven from the White Sands Missile Range to survey the northern sky and two from Woomera, Australia, to survey the southern sky. Each rocket carried a small 16.5-cm cryogenically-cooled telescope, and 90% of the sky was surveyed at the wavelength of 11 μm. As a result of the survey, over 2000 objects were catalogued. A program to identify these infrared sources was carried out at the 30-inch infrared telescope at *O'Brian Observatory* outside of Minneapolis and the 1.5-m infrared telescope at Mt. Lemmon in Arizona by Ed Ney of the University of Minnesota and Mike Merrill of the University of California at San Diego. One of the objects discovered in this ground-based identification program was object number 2688 in the AFGL catalogue. It was identified with a nebulous object and named the Egg Nebula by Mike Merrill based on its appearance in the *Palomar Observatory Sky Survey* plates. This is somewhat of a misnomer because with pictures taken at higher resolution, the object does not really look like an egg but has two separate lobes (Fig. 13.1).

At that time, it was not clear whether the Egg was a young or old star. In fact, in the discovery paper, Ed Ney debated between the interpretations that it was a newborn star in the process of evolving to the main sequence, or an old star that had evolved from the main sequence. At the suggestion of Roberta Humphreys of the University of Minnesota, David Crampton and Anne Cowley took

Figure 13.1.
[opposite] The *HST* WFPC2 image of the Egg Nebula. The two beams define the openings of the polar cone where light escapes. The dark lane between the two lobes is the dust torus. The central star is obscured behind this dark line.

an optical spectrum with the 1.8-m telescope at the *Dominion Astrophysical Observatory*. They classified the Egg Nebula as spectral type F5 Ia, meaning that it has a surface temperature of about 6000 °C and is very luminous. According to conventional wisdom, this makes the Egg Nebula a supergiant of very high mass (greater than ten times the mass of the Sun). However, the detection of absorption features due to the C_2 and C_3 molecules did not fit this interpretation. As we found out later, these spectral characteristics of the Egg Nebula would help us identify other proto-planetary nebulae. History had unknowingly been made: the first proto-planetary nebula had been discovered.

In order to find further examples of proto-planetary nebulae, a deeper survey of the infrared sky was needed. The *Infrared Astronomy Satellite* (*IRAS*), a joint mission between the US, the UK, and the Netherlands, promised to survey the entire sky in the mid-infrared and had the potential to reveal stars hidden in dust cocoons. In January 1982, during a workshop on stellar evolution held at the *Kitt Peak National Observatory*, Mike Merrill burst into the room and announced to the participants that *IRAS* had been successfully launched. Since I was giving a talk on the formation of planetary nebulae and the predicted properties of proto-planetary nebulae at the workshop, I was particularly excited about the news. A successful *IRAS* mission would allow us to mount a comprehensive search for proto-planetary nebulae. In the following 10 months before it ran out of coolants, the *IRAS* satellite surveyed 97% of the entire sky and catalogued over 250 000 sources. The mission was a tremendous success.

Although the *IRAS* survey gave us a starting point, finding an expected 100 or so proto-planetary nebulae among 250 000 potential candidates was still an insurmountable task. We needed to narrow down the search even more. I reasoned that the way to find proto-planetary nebulae was to search for far infrared objects that had infrared colors intermediate between the oldest

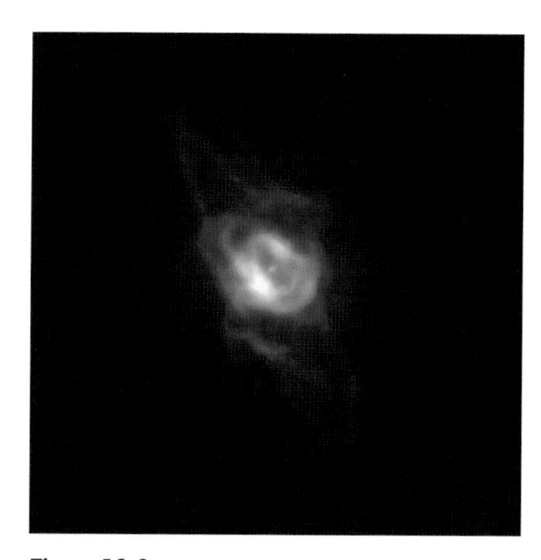

Figure 13.2.
M1-61, a young planetary nebula which appears stellar when observed with ground-based telescopes, was spatially resolved in our *VLA* survey (see Chapter 4). Hundreds of planetary nebulae remained unresolved optically until the launch of *HST*. This image of M1-61 was obtained with *HST* WFPC2.

asymptotic giant branch stars and the youngest planetary nebulae. In order to do this, we had to first identify a sample of young planetary nebulae and see what they looked like in the infrared.

Planetary nebulae are classified as such based on their spectrum – that is, their strong emission lines. In spite of their name, many of the entries in the planetary nebulae catalogue have a stellar appearance because the angular sizes of their nebulae are too small. Ground-based optical telescopes are limited by the Earth's atmosphere to about 1 arcsec in angular resolution. However, the *VLA* has about ten times the resolution and is capable of imaging very small (and likely young) nebulae (see Chapter 4). With the *VLA*, we identified a large number of very young planetary nebulae (Fig. 13.2).

Armed with this sample, we went on to compare the infrared colors measured by *IRAS* of these young planetary nebulae with those of the evolved asymptotic giant branch stars. To our amazement, we found a gap in the colors between these two groups. Since asymptotic giant branch stars are expected to evolve to planetary nebulae, objects that occupy this gap must be objects in the transition phase, or proto-planetary nebulae.

By searching through the *IRAS* catalogue for objects with colors that fit into this gap, Kevin Volk (University of Calgary), Bruce Hrivnak (Valparaiso University) and I selected a list of candidates. Since the *IRAS* positions were not very accurate, the first step was to identify their visual counterparts with a ground-based telescope. For several years, we carried out this program on the *Canada–France–Hawaii Telescope* (*CFHT*) and the *United Kingdom Infrared Telescope* (*UKIRT*) in Hawaii (Fig. 13.3) and the *Cerro Tololo InterAmerican Observatory* in Chile. In the plane of the Milky Way, where the density of stars is very high, there could be dozens or even hundred of stars near the position of *IRAS* sources. The only foolproof way to identify them was to search around the *IRAS* position using a mid-infrared detector. Once the optical counterpart of the *IRAS* source was identified, we proceeded to make infrared measurements. Since the stars in this transition phase are still evolving even when they are wrapped inside the dust cocoons, we had to probe through the cocoons and determine the evolutionary status of the stars. This is analogous to performing a prenatal ultrasound during pregnancy. Our objective – to find objects which were just emerging from their cocoons – seemed elusive. After many observations, there were still no clear signs of proto-planetary nebulae.

Figure 13.3.
The summit of Mauna Kea in winter. In spite of Hawaii's reputation for being a tropical paradise, the 4200-m summit can be cold and covered with snow. The telescope in the foreground is the *Canada–France–Hawaii Telescope*, the one immediately behind it is the University of Hawaii 2.2-m telescope, and the one with a metal dome is the 3.8-m *United Kingdom Infrared Telescope* (credit: J.-C. Cuillandre, Canda–France–Hawaii Telescope Corporation).

Then on May 26, 1986, Hrivnak and I were observing at the *CFHT* when we turned the telescope to *IRAS* 18095+2704, an infrared source in Hercules. Since most of the *IRAS* sources have faint visible-light counterparts, we turned up the gain on the monitor. When we scanned through the position, a bright star zapped into view, almost burning out the screen! This *IRAS* source was a 10th-magnitude star – bright, but not bright enough to be included in any existing star catalogue. Subsequent spectroscopic observations by Hrivnak identified it as spectral type F3 Ib, definitely not an evolved red giant or a young star but a luminous star of intermediate temperature (Fig. 13.4).

We determined that *IRAS* 18095+2704 had just emerged from its dust cocoon, and was therefore almost certainly a proto-planetary nebula. As the star stopped losing mass at the end of the asymptotic giant branch phase, we reasoned, its dust shell gradually dispersed with time. Eventually the star's light could be seen through the dust shell. By comparing the observational results with theoretical calculations by Kevin Volk, we estimated that *IRAS* 18095+2704 had left the asymptotic giant branch about 300 years ago.

Many similar discoveries soon followed. We have since found more than two dozen other yellow (spectral classes F or G)

Figure 13.4.
The bright star in the middle of this chart is *IRAS* 18095+2704, one of the first proto-planetary nebulae to be found. Based on the optical image alone, we would never know that it is such an interesting object.

supergiants with molecular absorption lines just like the Egg Nebula's. When the spectra of these stars are compared to the standard spectra in a stellar atlas, they are classified as belonging to a very high-luminosity class. However, this does not mean that they are supergiants evolved from very massive stars, as a strict interpretation of the 'rules' would imply. In fact, they are only a few thousand times brighter than the Sun, rather than 30 000 or 40 000 times brighter, as true supergiants are. When the 'rules' were formulated by William Morgan and Philip Keenan of *Yerkes Observatory* in 1943, these strange stars were unknown! And far from being massive objects, proto-planetary nebulae have only a fraction of the Sun's mass. When they were on the main sequence, these stars may have had several times the mass of the Sun, but most of it was shed during the preceding asymptotic giant branch phase.

While proto-planetary nebulae have very similar infrared characteristics, they differ greatly from one another in their visual brightness. Some are visually bright, and some are faint. Bruce Hrivnak and I were initially puzzled by this, but we realized it could be explained by viewing angle. If the dust is not symmetrically distributed around the star, a proto-planetary nebula viewed through a thinner part of the shell (face-on) will be visually bright. In contrast, if starlight has to pass through a lot of equatorial dust and suffer many magnitudes of extinction, the edge-on proto-planetary nebula will appear faint at visible wavelengths. This led us to postulate that proto-planetary nebulae are not round and symmetric. But without an acutal picture, we had no proof that this was indeed the case.

With the discovery of proto-planetary nebulae, we have at last found the last missing link in stellar evolution. The fact that they are short lived and rare makes the discoveries even more satisfying. These successes also brought new challenges. The first question on our mind was what do they look like? A clear answer to this question would come with observations from the *HST*.

14 Stellar metamorphosis

The hypothesis that proto-planetary nebulae may have asymmetric structures suggests that a planetary nebula may break out of its cocoon along certain preferred directions. The shell will be thin on the sides that are broken and remain thick on the sides that are intact. In order to test this idea, it would be desirable to take actual pictures of the nebulae. But unlike planetary nebulae, which are made up of glowing ionized gas, proto-planetary nebulae are not ionized and therefore do not actually radiate at visible wavelengths. Whatever brightness they have comes from starlight reflected off the surrounding dust. A bright central star typically outshines the small, faint nebula.

Using this reasoning, the best bet to find the surrounding nebulosity was from the faint, edge-on systems. In 1991, we attempted to do this using the *CFHT*'s high-resolution camera, which at the time could produce the highest image quality of any ground-based telescope. Two of our proto-planetary candidates, *IRAS* 17150-3224 and *IRAS* 17441-2411, showed unmistakable nebulosities with this instrument. In each case, the nebula is made up of two lobes with a dark dust lane in the middle. The stars, hidden behind the dust lanes, shine through polar gaps and illuminate the dusty reflection lobes on either side. Both show remarkable resemblance to the Egg Nebula, confirming that the Egg Nebula is not unique.

When I first presented the pictures of these two proto-planetary nebulae at a conference organized by Dimitar Sasselov at *Harvard*

Observatory in 1992, they generated a lot of interest because they showed conclusively that the morphological asymmetry seen in planetary nebulae develops very early. The debate of 'nature' vs 'nurture' had tilted in favor of the former.

Given the importance of the question, it was desirable to have better pictures of these two objects. Since there is no ground-based telescope that can give a better image, the next logical step was to use the *HST*. We observed these two nebulae with the *HST* Wide Field Planetary Camera 2 (WFPC2) in March 1997 and their images are shown in Figs. 14.1 and 14.2. In each case, we can see a series of concentric rings, representing the 'puffs' given off by the star in the last few thousand years of its life. After a number of these 'puffs', the star is wrapped inside a cocoon. In the *HST* images we are seeing the first indication that the nebulae are emerging from their cocoons, like butterflies undergoing metamorphosis.

We named the two nebulae the Cotton Candy Nebula and the Silkworm Nebula respectively. While 'Cotton Candy' gives a fair description of the first nebula's texture, the name Silkworm is particularly appropriate because it gives the correct scientific connotation.

We were very much encouraged by these results. The *HST* observations confirmed that our technique of seeking proto-planetary nebulae from infrared bright *IRAS* sources was working. It is amazing that small, faint, innocuous-looking objects can turn out to be beautiful nebulae! In 1998 and 1999, Bruce Hrivnak and I followed up a number of proto-planetary nebulae candidates that were only marginally resolved at the *CFHT* with the *HST* with similar success. Just compare the images of *IRAS* 16594-4656 and 17245-3951 in the Digital Sky Survey (Fig. 14.3) and the *HST* WFPC2 images of these two objects in Figs. 14.4 and 14.5. *IRAS* 17245-3951 seems to be a smaller version of the Cotton Candy nebula with two bipolar lobes lying on the plane of the sky. However, *IRAS* 16594-4656 has a flower-like morphology with

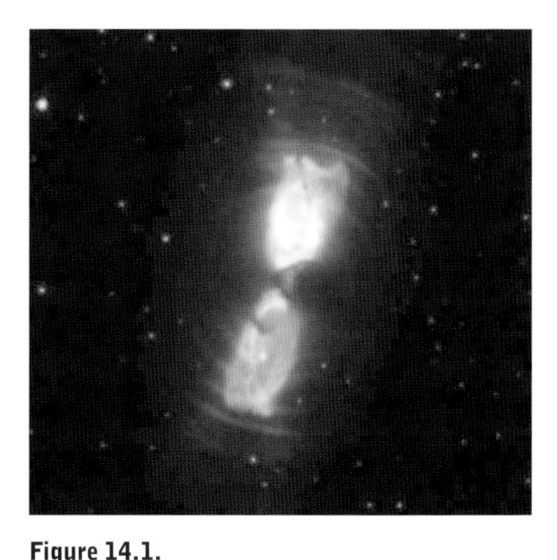

Figure 14.1.
The Cotton Candy Nebula as imaged by the *HST*. This proto-planetary nebula in the constellation of Scorpius bears a strong resemblance to the Egg Nebula (Fig. 13.1). As in the Egg Nebula, a series of concentric arcs can be seen on both lobes. The nebulosities of proto-planetary nebulae such as the Cotton Candy are due to starlight reflected from dust particles, not emission lines as in planetary nebulae.

Figure 14.2.
The Silkworm Nebula in Sagittarius resembles a moth emerging from its cocoon; hence its name.

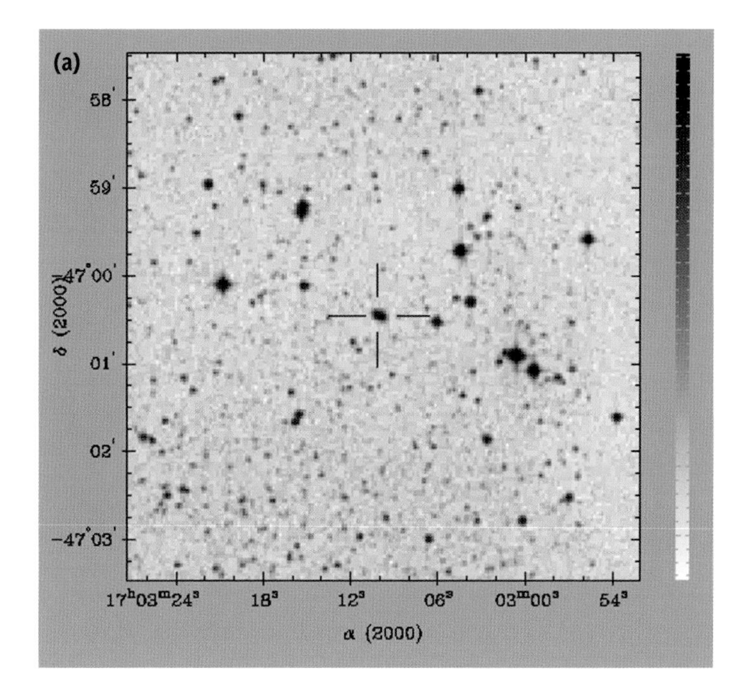

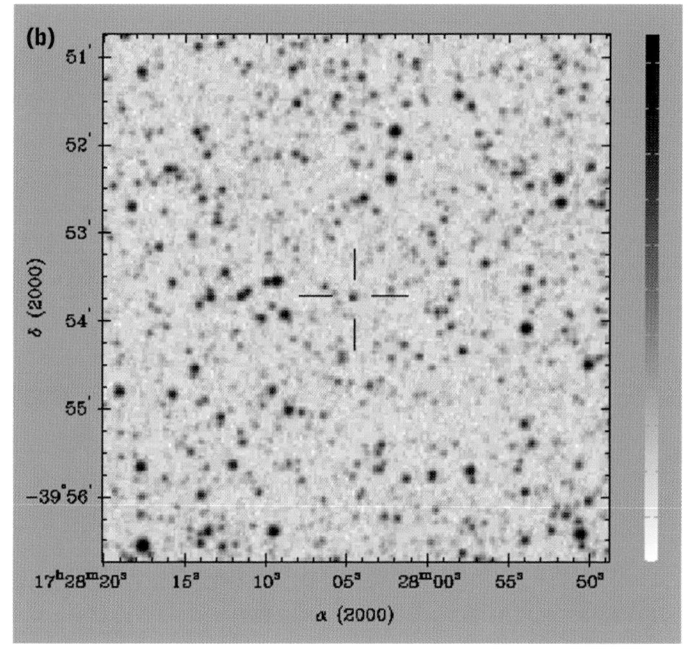

Figure 14.3.
The images of *IRAS* 16594-4656 (a) and 17245-3951 (b) in the *Digital Sky Survey* plates. From these pictures, one would never guess that the small dots in the middle of the two fields in fact have beautiful nebulous structures as seen in Figs. 14.4 and 14.5.

petals extending outwards. It was immediately clear to us that we were not viewing 16594-4656 edge-on, but at an angle. Again from their apparent shapes, we named these two objects the Walnut Nebula and the Water Lily Nebula.

Among the most remarkable images we obtained was *IRAS* 17106-3046, which we named the Spindle Nebula (Fig. 14.6). For the first time, the equatorial disk is clearly outlined in scattered light. In contrast to the Egg, Cotton Candy and Silkworm, the two lobes of the Spindle Nebula have pointed ends, suggesting that the fast outflow is just breaking out of the cocoon. These images were analyzed by my graduate student Kate Su, who was able to derive the geometry and quantitative structures of the nebulae.

A number of reflection nebulae believed to be proto-planetary nebulae were also imaged with the *HST* by Raghvendra Sahai of the Jet Propulsion Laboratory, Valentin Bujarrabal of Observatorio Astronómico Nacional of Spain, and Albert Zijlstra of Manchester

Institute of Science and Technology. The Boomerang Nebula, shown in Fig. 14.7, has two fan-shaped lobes and a very tight waist. Similar to the Cotton Candy and the Silkworm is M1-92 (also known as 'Minkowki's footprint'), where one lobe is significantly brighter than the other (Fig. 14.8). This is because the brighter lobe is tilted towards us and the fainter lobe is obscured by a dust torus around the waist. The dust torus, in the form of a dark lane, is clearly seen in the images of He 3-401 (Fig. 14.9), Roberts 22 (Fig. 14.10) and Mz-3 (Fig. 14.11), bipolar nebulae being viewed sideways. Unlike the fan-shaped lobes that we see so often in planetary nebulae, the object He 3-401 has a remarkable cylindrical shape. Such cylindrical lobes are also present in Mz-3; in addition, there is a pair of bright round lobes near the central star.

The discovery of such a cylindrical shape surprised astronomers because such shapes suggest that the fast wind may be directional. The cocoon is broken not only because it is thinner in the polar directions, but probably also because the fast wind causing breaking is preferentially directed. We may be seeing the future of He 3-401 in the picture of IC 4406 shown in Fig. 14.12, a planetary nebula in the southern constellation of Lupus. The cylindrical lobes are still present, but the central torus is now ionized and shows up as a ring in the center.

In February 1997, astronauts on the space shuttle *Discovery* installed the *Near Infrared Camera and Multi Object Spectrometer* (*NICMOS*) on the *HST*. *NICMOS* is made up of an array of HgCdTe infrared detectors which can take infrared images in the wavelength range between 0.8 and 2.5 μm (Figs. 14.13–14.15). These detectors have to operate at very low temperatures and are kept cold inside a thermally insulated container (like a thermos bottle) containing frozen nitrogen ice.

One of the unique properties of *NICMOS* is its capability to isolate the presence of molecular hydrogen. Molecular hydrogen is useful because it records the cracking sound of the shell. We can

Figure 14.4.
This *HST* picture of the Walnut Nebula in the constellation of Scorpius shows clearly two bipolar lobes separated by a dark dust lane and surrounded by a outer halo.

Figure 14.5.
The Water Lily Nebula in the constellation of Ara. The outer petals of the nebula extend to 6.4 arcsec from the central star.

Figure 14.6.
[above] The Spindle Nebula shows a collimated bipolar outflow emerging from the center of a visible disk. The two lobes appear brighter along their edges, suggesting that they are hollow cavities reflecting light from their walls. The cocoon is clearly defined by the 7 × 8 arcsec halo.

actually hear the sounds of breakout from the cocoon. With the *NICMOS* picture of the Cotton Candy nebula shown in Fig. 14.16, we are witnessing the birth of a planetary nebula!

The transformation from star to planetary nebula is one of the most remarkable miracles in astronomy. Near the end of its life, a star begins to eject gas and dust from its surface and manages to enclose itself in a cocoon in only a few hundred thousand years. For many stars, the cocoons can be so thick that the stars become invisible from view. The only way we can tell that they are there is by detecting the heat that they give out using infrared telescopes.

Figure 14.7.
[right] The Boomerang Nebula in Centaurus was discovered by Gary Wegner of the *South Astronomical Observatory* when he examined one of the *European Southern Observatory* (*ESO*) sky survey plates and noticed its peculiar 'bow-tie' shape. This *HST* WFPC2 picture shows its two lobes fanning out from a remarkably tight waist.

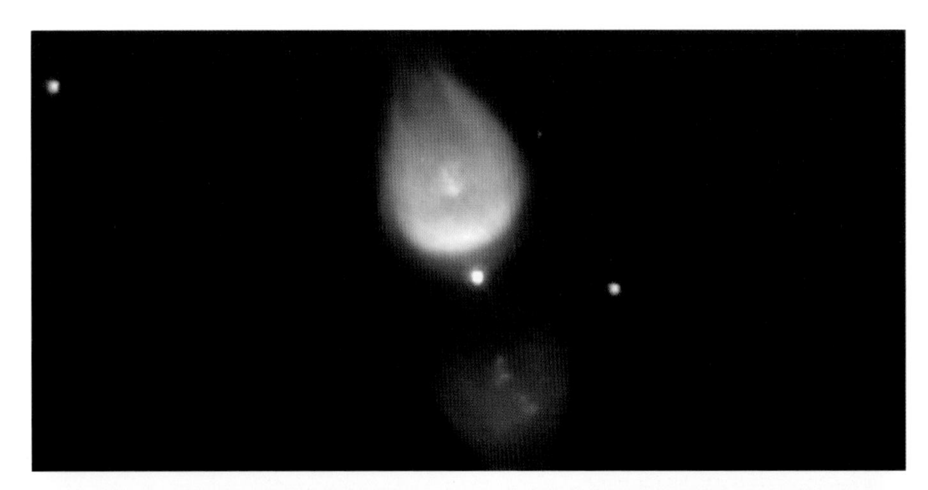

Figure 14.8.
M1-92, also known as Minkowski's footprint, is a reflection nebula with a 20 000 °C central star and therefore a possible proto-planetary nebula candidate. Its two bipolar lobes are separated by a dust torus. The northwest lobe is much brighter than the southeast lobe, probably because the former is tilted towards the observer and the latter is partly obscured by the equatorial dust.

Figure 14.9.
He 3-401 was first discovered as a peculiar star showing emission lines by Karl Henize during his survey of the southern skies. His observations were made from Bloemfontein, South Africa, using a 10-inch telescope originally from Mt. Wilson. Between 1949 and 1951, he surveyed the entire southern sky south of declination −25 ° in search of emission-line stars. He 3-401 is one of the 1929 objects in his 3rd catalog published in 1976. Its associated nebulosity was discovered by David Allen in 1977 with the 1.88-m telescope at *Mount Stromlo Observatory* in Australia. Although it shows emission lines in its optical spectrum, its central star is too cool (about 22 000 °C) to be considered a genuine planetary nebula. The detection of strong infrared dust emission by the *IRAS* satellite led to the suggestion that it is a proto-planetary nebula. This *HST* WFPC2 image shows that He 3-401 is a bipolar reflection nebula with a cylindrical shape. Its two lobes, each about 15 arcsec long and 2–2.5 arcsec wide, are separated by an equatorial dust torus. The almost tube-like nebulosity suggests that the fast outflow is highly collimated.

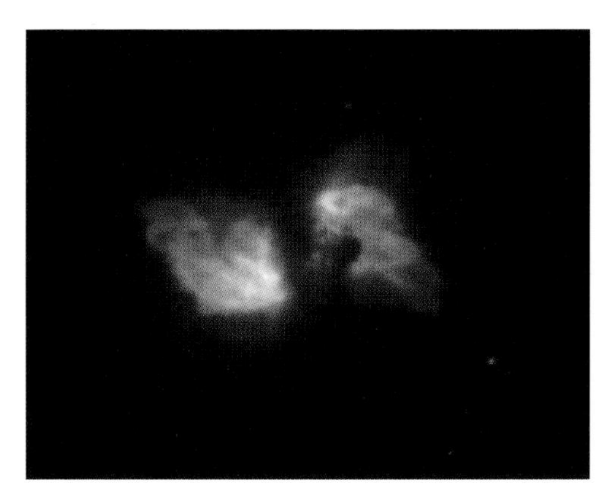

Figure 14.10.
[above] Roberts 22 is another reflection nebula with a cool (10 000 °C) central star. This *HST* picture shows two bright bipolar lobes shaped like butterfly wings. The dark lane between the two lobes is probably due to an equatorial dust torus. The lobes are surrounded by a fainter halo.

Figure 14.11.
[right] Although Mz-3 was discovered by Don Menzel in 1922, its nature was never understood. Because of its nebulous structure, it was included in the 1967 planetary nebulae catalogue of Perek and Kohoutek. However, its spectrum is inconsistent with that of a planetary nebula. In 1978, Martin Cohen of University of California, Berkeley, estimated its central star temperature to be 32 000 °C. This *HST* picture of Mz-3 shows a pair of bipolar lobes with long skirts.

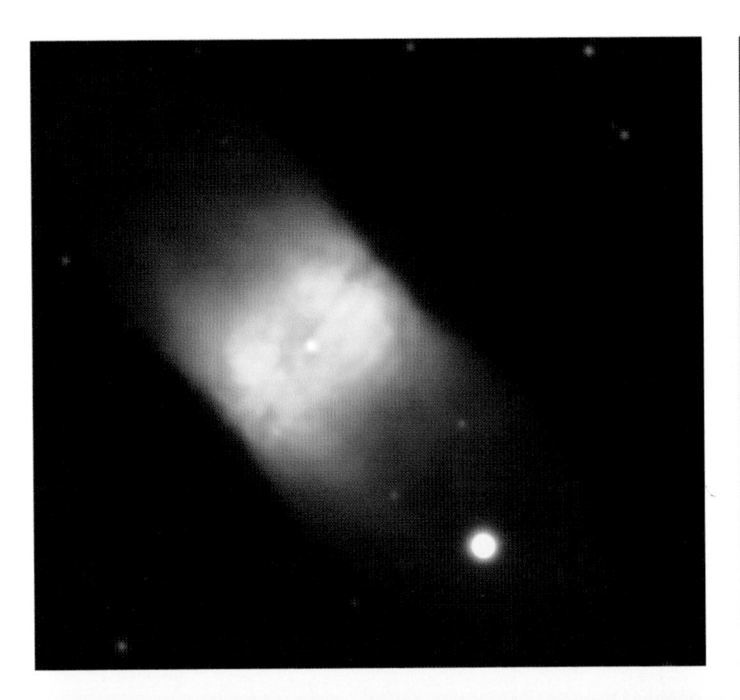

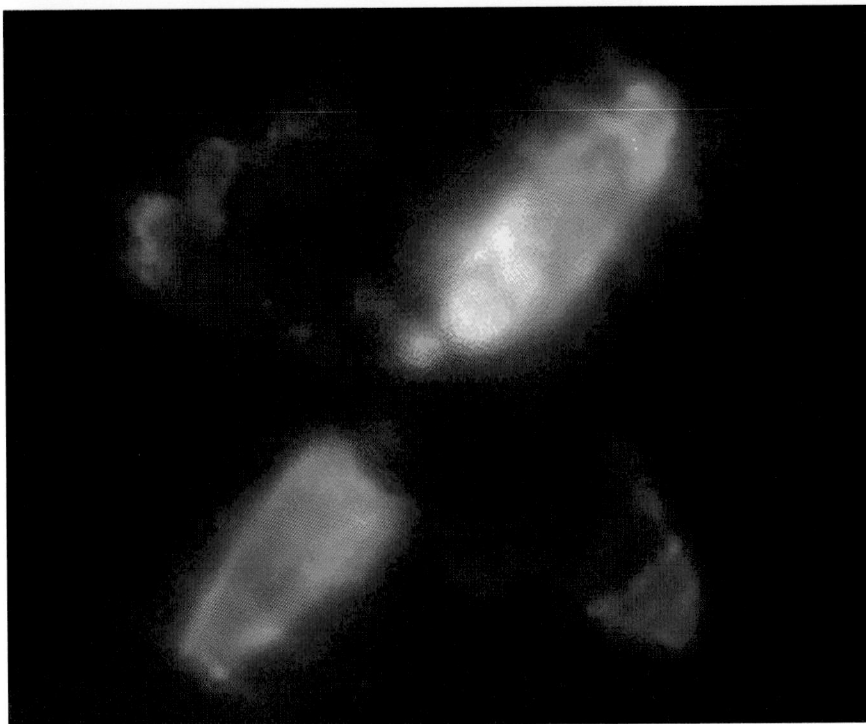

Figure 14.12.
[above left] This picture of IC 4406 taken at the *European Southern Observatory Very Large Telescope* shows a central ring and cylindrical lobes (credit: ESO).

Figure 14.13.
[above right] The power of *NICMOS* is illustrated in this infrared picture of the Trapezium star formation region in Orion. Many of the newborn stars in the Trapezium region are obscured by dust, and are only revealed in infrared light (credit: Eric Young).

Figure 14.14.
[left] *HST NICMOS* image of the Egg Nebula with the molecular hydrogen emission shown in red (credit: R. Sahai).

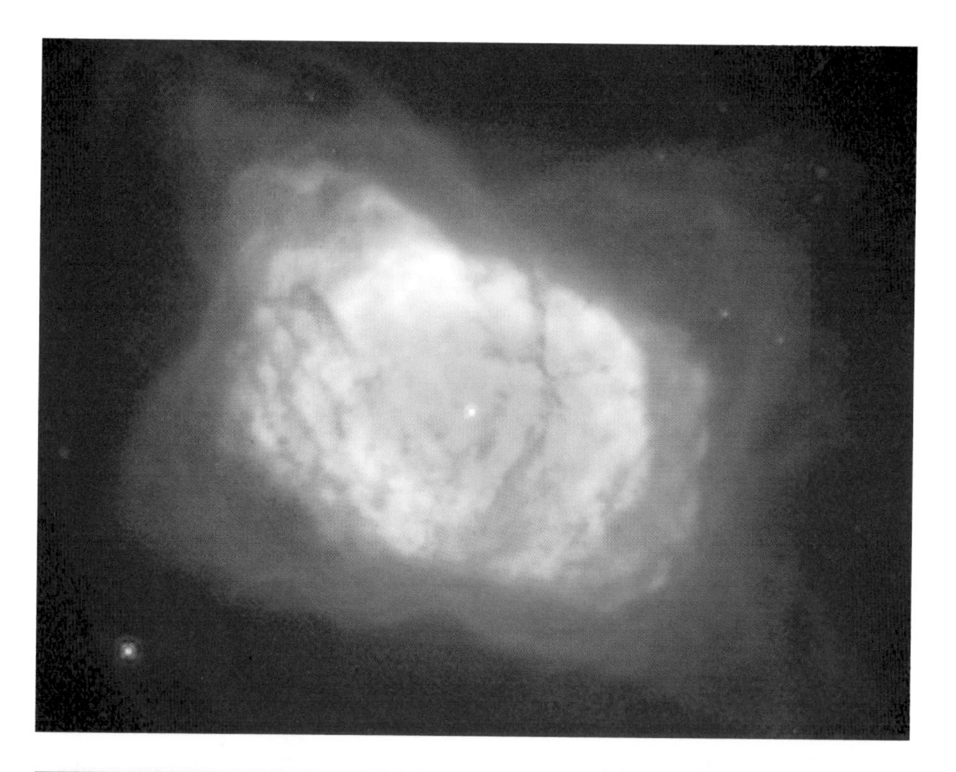

Figure 14.15.
A combined WFPC2 and *NICMOS* image of NGC 7027. The familiar ring-shaped ionized nebula is seen in white, whereas the pink color is due to emission by molecular hydrogen. The amount of mass in the ionized nebula is about 0.02 solar masses, much less than the estimated 3 solar masses contained in the much larger and optically invisible molecular envelope (see Chapter 10). The molecular hydrogen traces the boundary between the ionized gas and the molecular envelope and is a clear demonstration of the process of interacting winds at work. We can see that the molecular hydrogen gas has an hourglass shape. This image was analyzed by Aditya Dayal and Bill Latter to show that NGC 7027 is in fact a bipolar nebula tilted at an angle of 45° from the plane of the sky (credit: W. B. Latter).

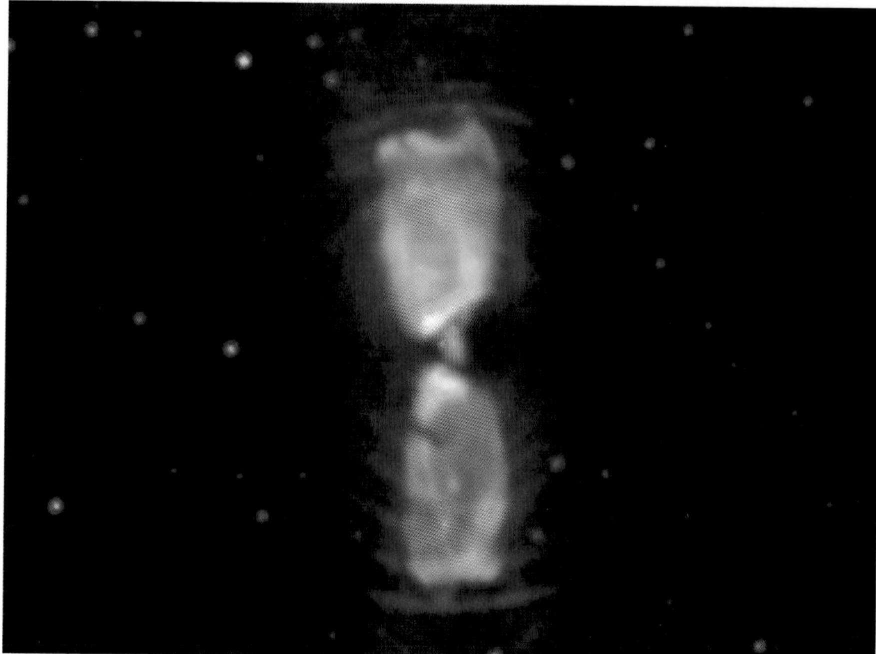

Figure 14.16.
Image of the Cotton Candy Nebula showing the regions of molecular hydrogen emission in red. The molecular hydrogen emission is generated as the result of high velocity winds ramming against the dust cocoon created during the preceding asymptotic giant branch evolutionary phase.

When they use up the material in their hydrogen envelope in creating this cocoon, a fast outflow develops and eventually breaks the shell at the two ends. A butterfly shape nebula begins to take form by sweeping up the gaseous material in the shell. In about one thousand years, the hidden star becomes hot and the ultraviolet light that it emits ionizes and lights up the surrounding material. A planetary nebula is born.

15 Unsolved mysteries

While we have made some advances in the understanding of the global structures of planetary nebulae, new questions have arisen as the result of the observations by the *HST*. The high-resolution capability of the *HST* allows small-scale local structures to be studied. For example, in the picture of NGC 6826 shown in Fig. 15.1, we see a pair of 'handles' (or ansae) on two opposing sides of the shell. These ansae appear in blue because they are particularly prominent in the neutral oxygen and ionized nitrogen lines. Although these microstructures are seen with unprecedented clarity with the *HST*, their existence has in fact been known for over 60 years. In the late 1930s, Lawrence Aller used the 36-inch Crossley reflector to observe a number of planetary nebulae. In particular, he noted that 'NGC 7009, a beautiful and interesting planetary, exhibits a spindle shape and a pair of faint extensions terminated by bright ansae'. NGC 7009 remains one of the best examples of ansae, as you can see in the *HST* picture in Fig. 15.2.

The search for microstructures was taken up by Bruce Balick of the University of Washington after CCD cameras became available. Using the *Kitt Peak National Observatory* 2.1-m and the *Palomar* 200-inch telescopes, Balick took CCD images and spectra of NGC 3242 (Fig. 15.3), NGC 7662 (Fig. 15.4), and others to study the ansae. He found that the ansae are expanding faster than the nebular shell, and he termed them Fast, Low-Ionization Emission Regions, or FLIERs. We can consider FLIERs as bullets shot out from the nebulae because they travel at five times the local sound speed (or

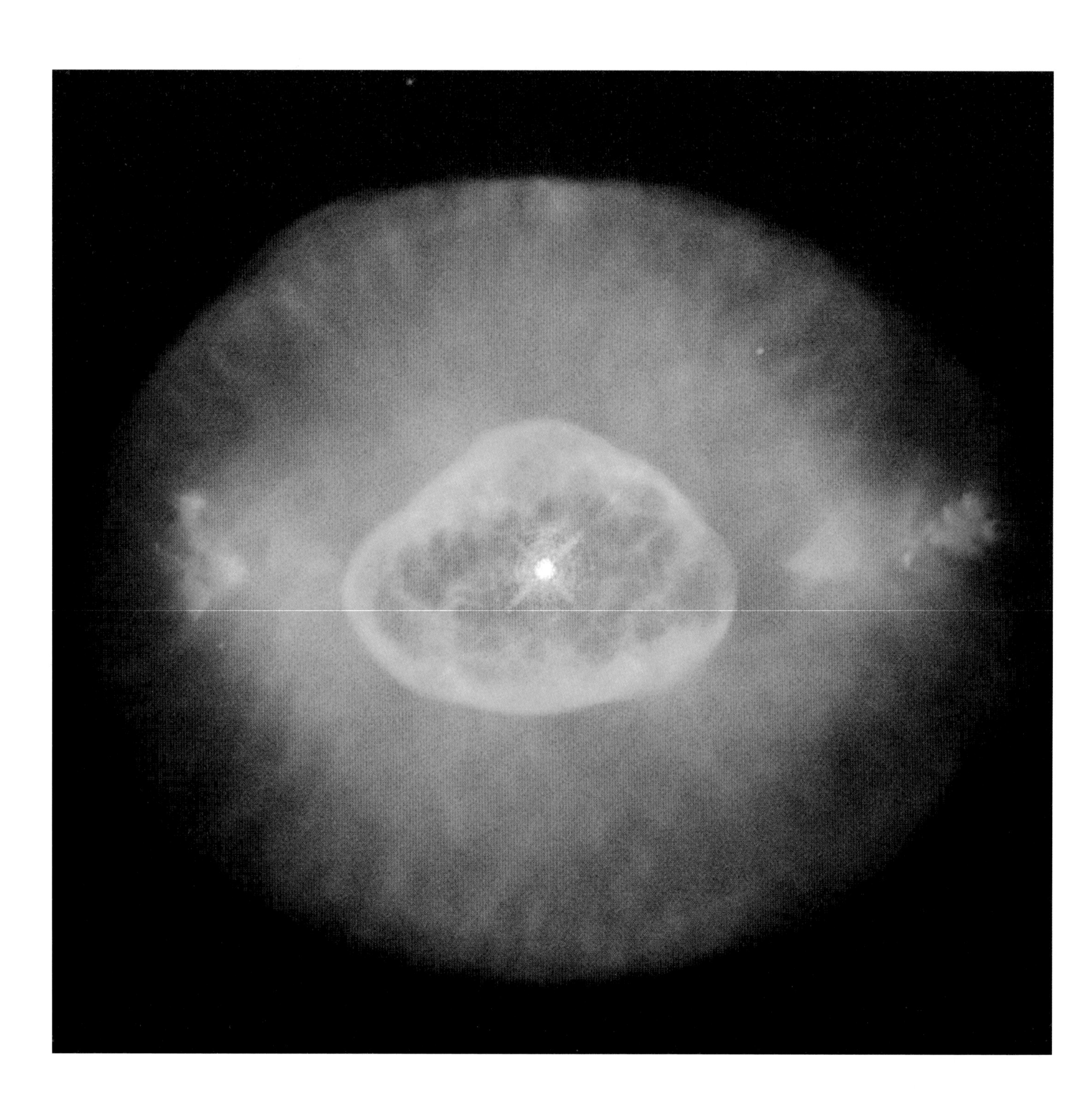

Figure 15.1.
[opposite] The richness and complexity of planetary nebulae is clearly illustrated in this *HST* image of NGC 6826. The central star is at the center of a slightly elongated nebular shell and an outer 'crown'. Two FLIERs can be seen near the edge of the outer shell.

Mach 5 in aerospace jargon). They are always found at approximately equal distances, but in opposite directions, from the central star. How and why planetary nebulae would shoot out such fast bullets remains a mystery to this day (Fig. 15.5).

As astronomers take better pictures with the *HST*, we run the risk of destroying the romantic images of objects that we grew accustomed to. The Eskimo Nebula (NGC 2392) is a case in point. When viewed through ground-based telescopes, it resembles a face surrounded by a fur parka. On January 10 and 11, 2000, after the successful third service mission of the *HST* by the space shuttle *Discovery* (Fig. 15.6), the staff of *Space Telescope Science Institute* took the image of the Eskimo Nebula shown in Fig. 15.7. The face of the Eskimo turns out to be filled with intersecting helical

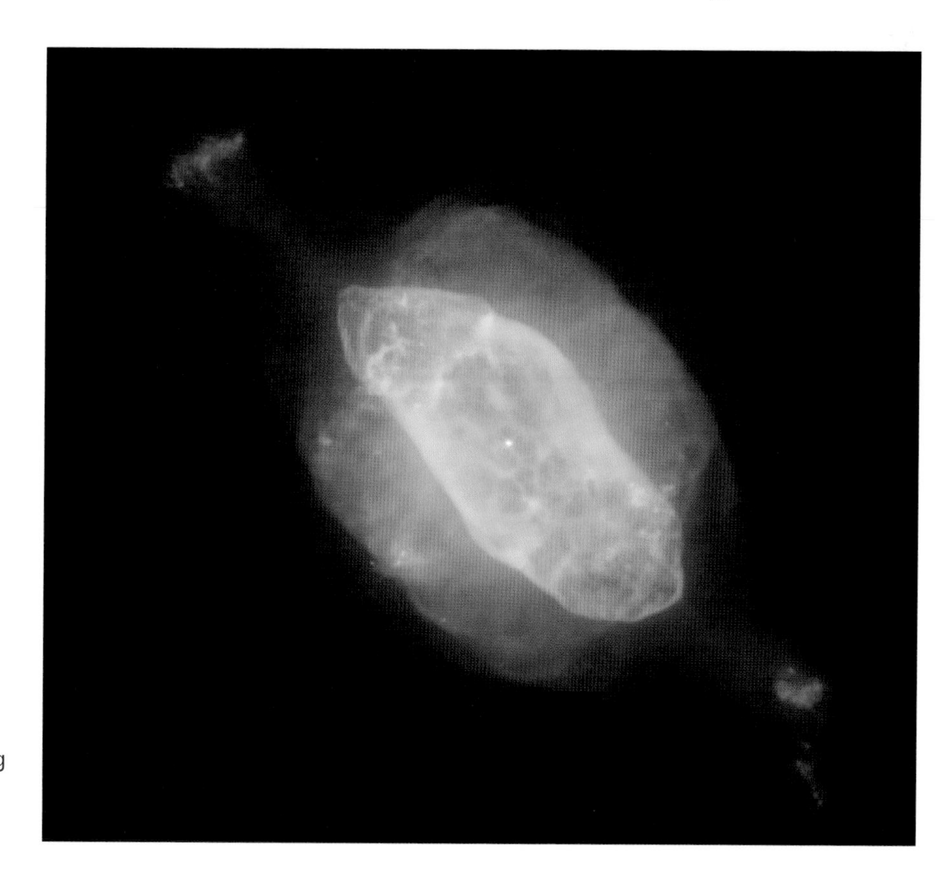

Figure 15.2.
The Saturn Nebula (NGC 7009) in Aquarius was discovered by William Herschel in September 1782 during his survey of the sky. This *HST* image of NGC 7009 shows a pair of FLIERs along the major axis of the planetary nebula. A 'crown' can also be seen outside the nebular shell.

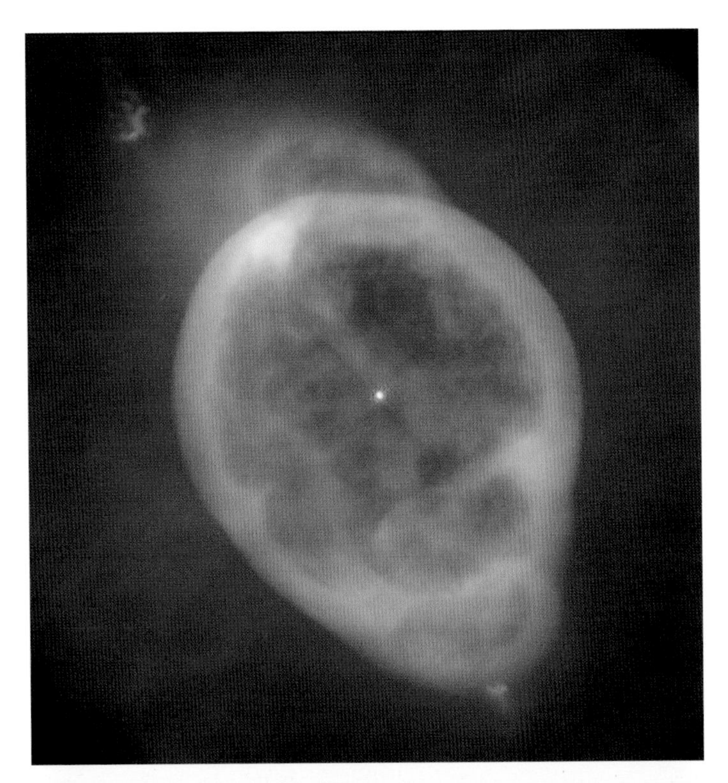

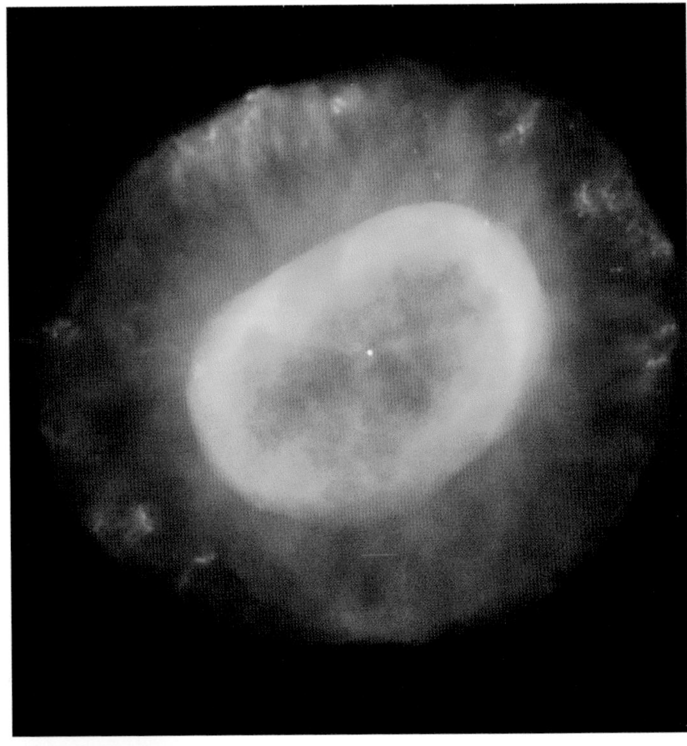

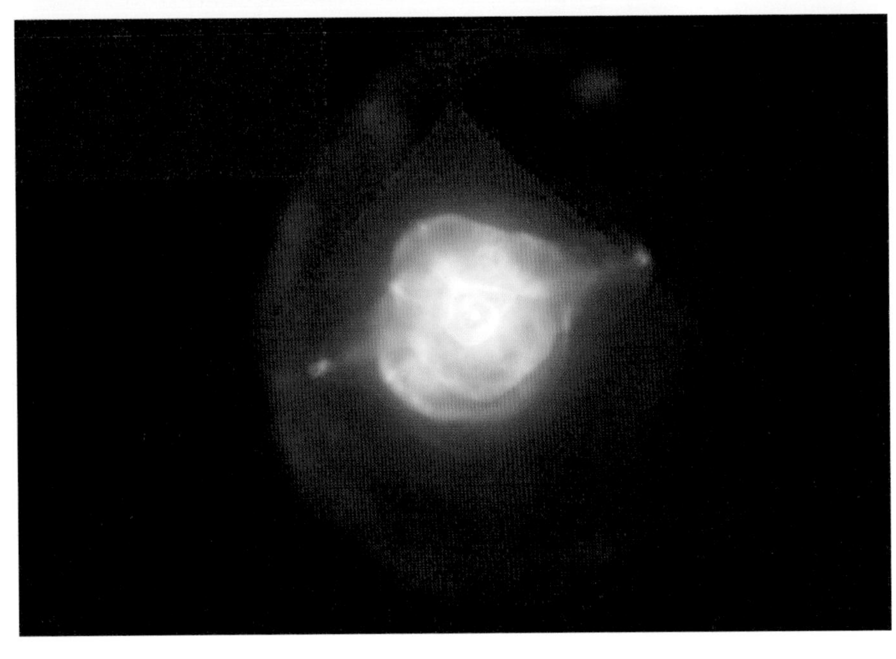

Figure 15.3.
[above left] A pair of FLIERs can be clearly seen outside the elliptical shell of NGC 3242.

Figure 15.4.
[above right] The elliptical planetary nebula NGC 7662 has a ring-shaped shell and an outer 'crown'. A pair of FLIERs as well as a dozen microstructures can be seen at the periphery of the outer crown along the major axis.

Figure 15.5.
[left] A pair of 'bullets' (each about 12 arcsec from the central star) can be seen in IC 4593. Trails of material can be seen connecting these 'bullets' with the inner shell, making IC 4593 one of the best examples of the existence of jet-like structures in planetary nebulae.

filaments, and the fur in the hat is a series of knots and jets emanating from the halo (Fig. 15.8).

This image of the Eskimo Nebula was modeled by John Blondin and Kazik Borkowski of North Carolina State University. They suggest that the 'fur hat' represents the equatorial enhanced slow wind and that the 'face' is the result of a fast wind blowing through the poles. These wind interactions were simulated with hydrodynamical calculations using the North Carolina Supercomputing Center. The results of the simulations show that the 'bubbles' become unstable after they grow to a certain size. The breakup of the bubbles is the origin of the filaments on the 'face' (Fig. 15.9).

Similar thin, radial filaments in the Helix Nebula were noticed by Walter Baade as early as the 1930s. This discovery was largely neglected until B.A. Vorontsov-Velyaminov of the Sternberg State Astronomical Institute of the former USSR brought it to the attention of the participants of the International Astronomical Union symposium on planetary nebulae in Tatranska Lomnica, Czechoslovakia, in 1967. He found that these filaments are pointing directly at the central star and suggested that they are gaseous streams ejected from the star. When Bob O'Dell of Rice University examined the outer regions of the Helix Nebula with the *HST*, he found that these filaments are resolved into numerous knots (Fig. 15.10). Seen in silhouette against the background of nebular emission, these knots appear to have a central dark core and a luminous cusp on the side facing the star. Since they resemble comets in our own solar system, O'Dell named them cometary knots. We are able to see such details because the Helix Nebula is so close to Earth. It is possible that other planetary nebulae, if examined by telescopes more powerful than the *HST*, could have similar microstructures (Fig. 15.11).

Another example of microstructure is a series of linear structures called 'jets' that can be traced back to the central star. The *HST*

Figure 15.6.
[opposite] After the fourth gyroscope failed, the *HST* stopped observing and was put in the safe mode in November 1999. On December 20, 1999, the space shuttle *Discovery* captured the telescope, replaced equipment, and performed maintenance upgrades. This picture shows the *HST* floating gracefully above the blue Earth after release from *Discovery's* robot arm (credit: NASA).

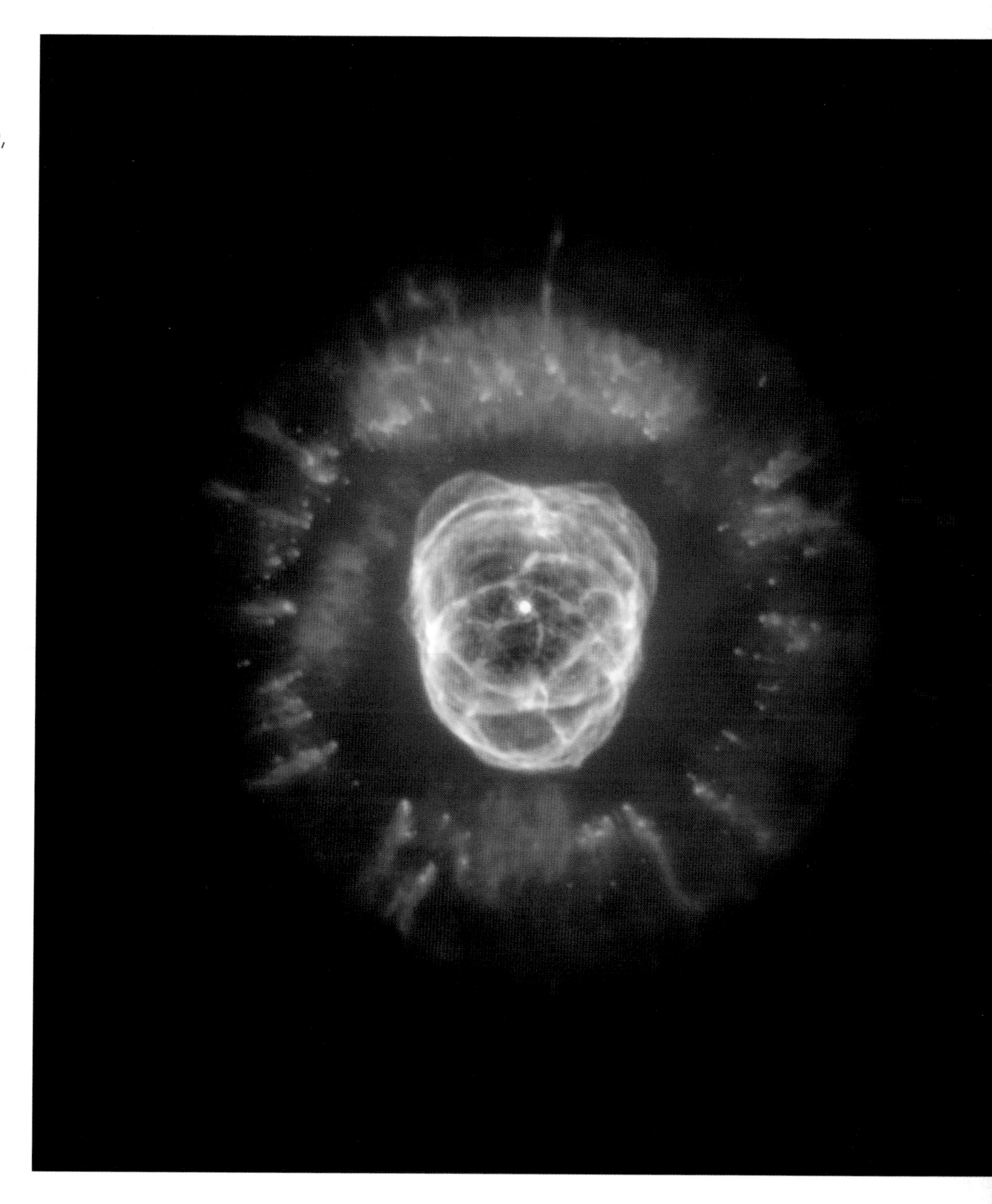

Figure 15.7.
[right] This picture of the Eskimo Nebula was one of the first images obtained with the *HST* Wide Field Planetary Camera 2 after the third service mission (credit: NASA and the Hubble Heritage Team, Space Telescope Science Institute).

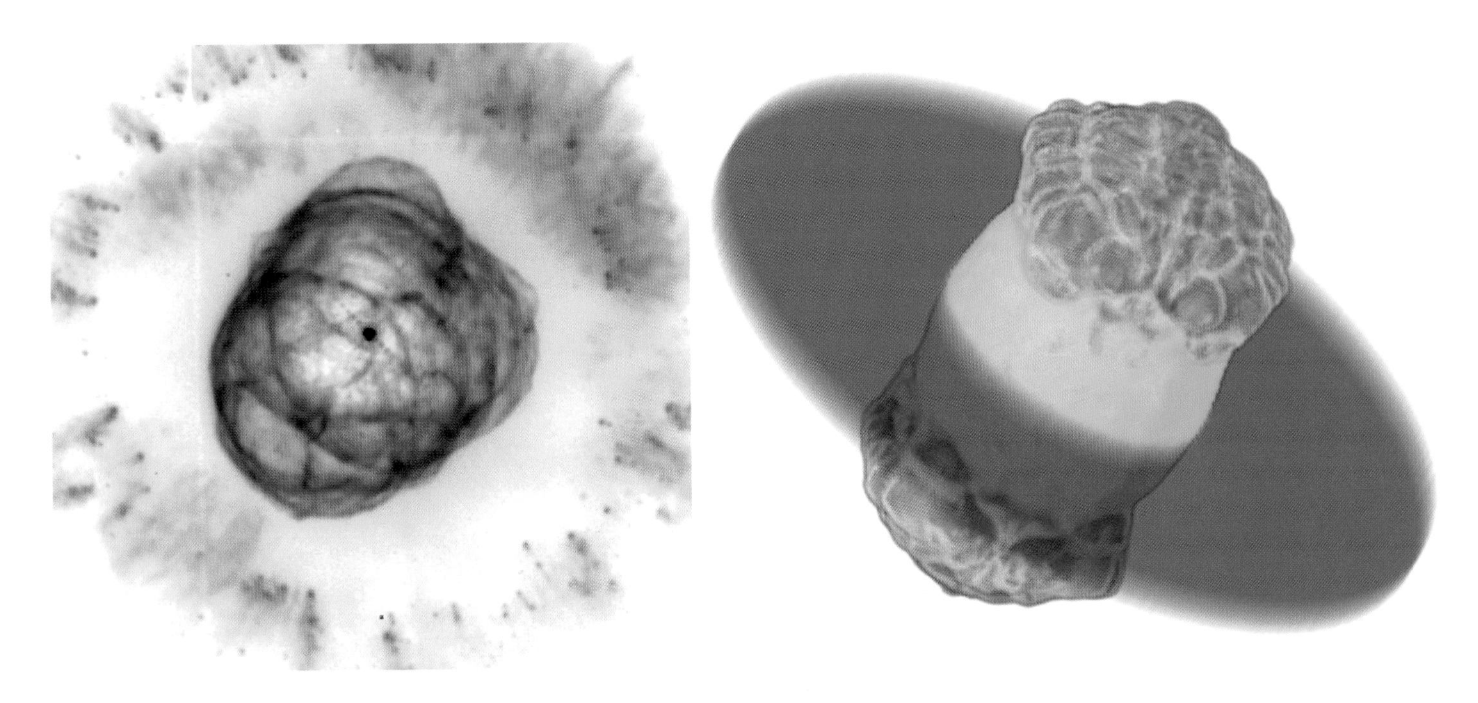

Figure 15.8.
[above left] A close up view of the Eskimo Nebula, showing the filamentary structure of the nebular shell.

Figure 15.9.
[above right] A computer model of the Eskimo Nebula showing the fast outflow bursting out of the equatorial disk. The disk is the remnant of the slow wind which was preferentially ejected in the equatorial directions. As a later-developed fast outflow breaks out from the polar directions, the shock creates numerous small bubbles. Our view to the Nebula is down the axis perpendicular to the disk. The disk forms the fur hat, and the projections of the small bubbles form the face of the Eskimo (credit: John Blondin, Kazimierz Borkowski and Patrick Harrington).

image of the Cat's Eye nebula revealed a number of faint lines that point toward the central star (Fig. 15.12). These 'jets' are probably ejected from the central star at high speed and have penetrated the nebular shell (see also Fig. 15.13).

In 1998, we proposed to use the *HST* to obtain optical images of a number of young, compact planetary nebulae that we had previously observed with the *VLA*. Although we expected the *HST* to reveal some faint outer structures not measurable by radio techniques, we were totally unprepared for the results we got. When my student Kate Su processed the data, she found more than one set of bipolar lobes in several of these nebulae. Figure 15.14 shows an example of such a multi-polar nebula. We had found a new species of butterfly with two pairs of wings!

One of the most bizarre of all planetary nebulae is NGC 2440 (Fig. 15.15). Minkowski described it in 1964 as 'an example of an object so complicated it defies description'. Its main structure is a pair of bipolar lobes extending over 1 arcmin in total size. In 1998,

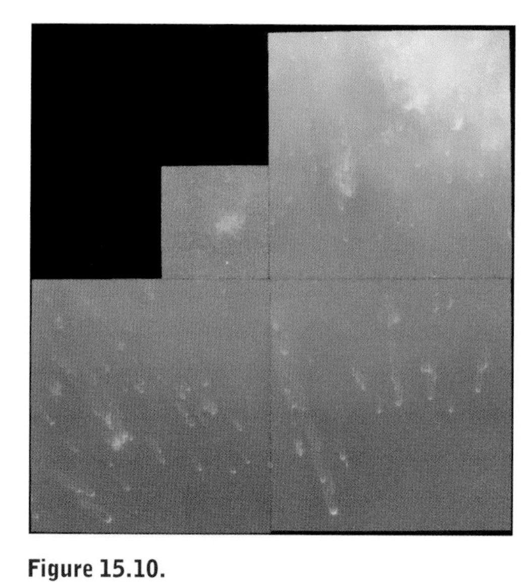

Figure 15.10.
[above] A close up view of the northern ring of the Helix Nebula. Because of the large size of the nebula, only a small fraction of the nebula is covered in this *HST* image. Bob O'Dell counted 313 knots in this image and he estimated that there are about 3500 knots in the entire Helix Nebula.

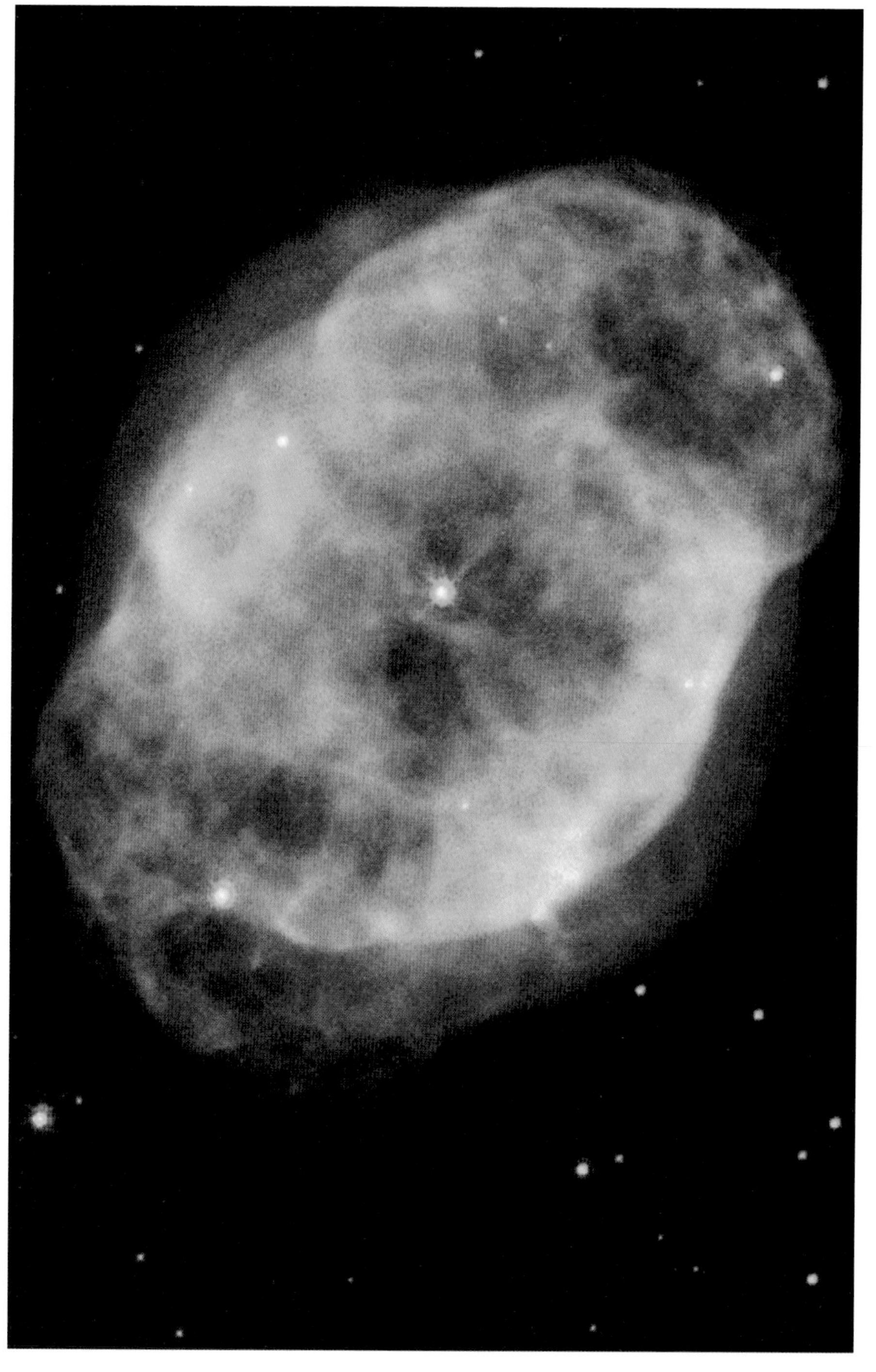

Figure 15.11.
[right] This *HST* image of IC 4663 clearly shows the filamentary structure of the nebular shell.

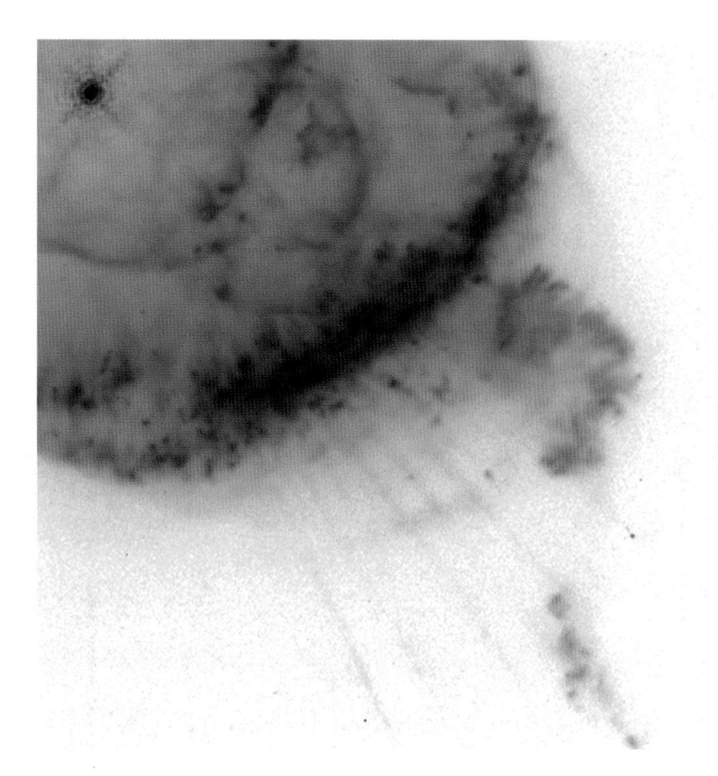

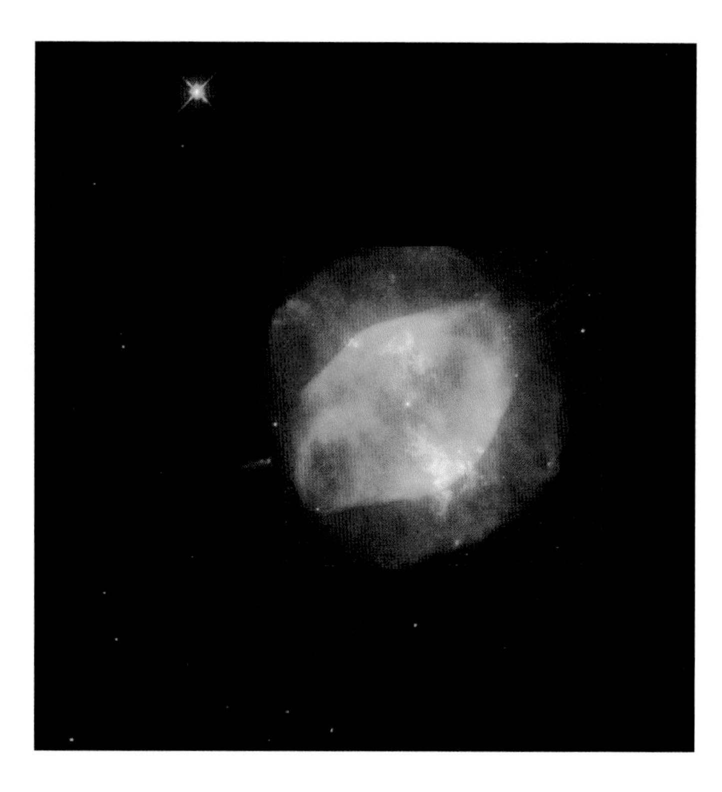

Figure 15.12.
[above left] A series of jets can be seen in this image of NGC 6543 taken in the light of singly ionized nitrogen.

Figure 15.13.
[above right] Two jet-like features (in red) can be seen emerging from the spherical halo surrounding the elliptical shell of NGC 7354.

Alberto López of Unversidal Nacional Autónoma de México took a deep image of this nebula and found that it has another pair of lobes aligned along a different axis! Were these two pairs of lobes ejected at different times, or were they ejected simultaneously but along different directions? We have no answer to this question.

The problems presented by the new *Hubble* images were brought home to the participants of the Second International Conference on Aspherical Planetary Nebulae at the Massachusetts Institute of Technology in August 1999. Astronomers were confronted by structures that defied conventional wisdom. One of the buzzwords of the conference was 'point-symmetric structure', a technical term referring to blobs and filaments on one side of the nebula being reproduced exactly upon reflection onto the other side. An example of such a structure is the letter 'S'. Back in the 1980s, we had already noticed in our radio survey that certain planetary nebulae

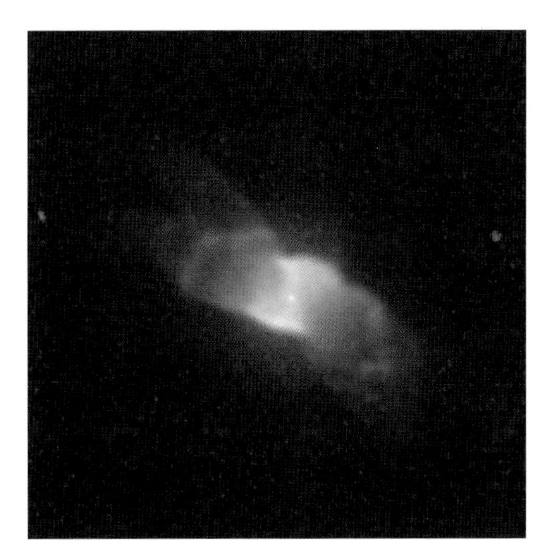

Figure 15.14.
Two sets of butterfly wings can be seen in this *HST* image of He 2-447. The existence of multi-wing butterflies was totally unexpected.

Figure 15.15.
NGC 2440 has two pairs of bipolar lobes oriented in different directions. Because of the limited field of view of the *HST* wide field camera, only portions of the lobes are displayed in this picture. A close-up view of this nebula is shown in Fig. 3.1.

have the 'S' shape. One example is K3-35, which my student Orla Aaquist named the Seahorse Nebula (Fig. 15.16). In 1995, Alberto López discovered large, faint, bipolar lobes outside the KjPn 8, a planetary nebula first discovered by M. Kazarian and E. Parsamian of *Burakan Astrophysical Observatory* in Armenia in 1971. Most interestingly, three pairs of knots can be seen, giving the impression that they are created by three ejection episodes (Fig. 15.17). López called them bipolar rotating episodic jets, or BRETs for short. He suggested that the point-symmetric knots were ejected by a wobbling nozzle in the core. Just imagine that your backyard sprinkler is not working too well, and the water pattern that it throws out is a BRET!

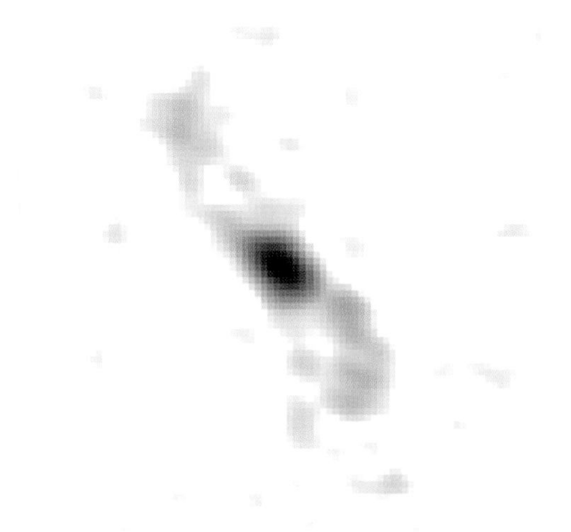

Figure 15.16.
This radio image of the Seahorse Nebula K3-35 taken with the *VLA* is one of the first examples of planetary nebulae having an 'S' shape.

A more dramatic 'S'-shape morphology is shown in the *HST* picture of NGC 5307 (Fig. 15.18), which upon casual inspection might have been mistaken for a spiral galaxy! If the S-shape is created by a wobbling sprinkler (see Figs. 15.19–15.21), then what is the nozzle? At the MIT conference, the participants were divided into two camps. One camp, headed by Noam Soker of the University of Haifa and Mario Livio of the *Space Telescope Science Institute*, believed that the nozzle emerges from disks created by material flowing to the central star from an unseen binary companion. Another camp, represented by Valentin Bujarrabal of the *Observatorio Astronomico Nacional, Ministerio de Fomento* in Spain argued that single stars, perhaps with the help of magnetic fields, can do the job. Others participants recognized the attractions of the binary model, but were afraid that accepting the idea would open a whole new can of worms. The binary model could generate so many scenarios that it would be difficult to test. Since science relies on quantitative predictions that can be checked by precise observations, an attractive idea is not equivalent to a good theory.

Another new, unexpected outcome of *HST* observations of planetary nebulae was the discovery of concentric arcs. These arcs were first seen in bipolar proto-planetary nebulae in 1998. A series

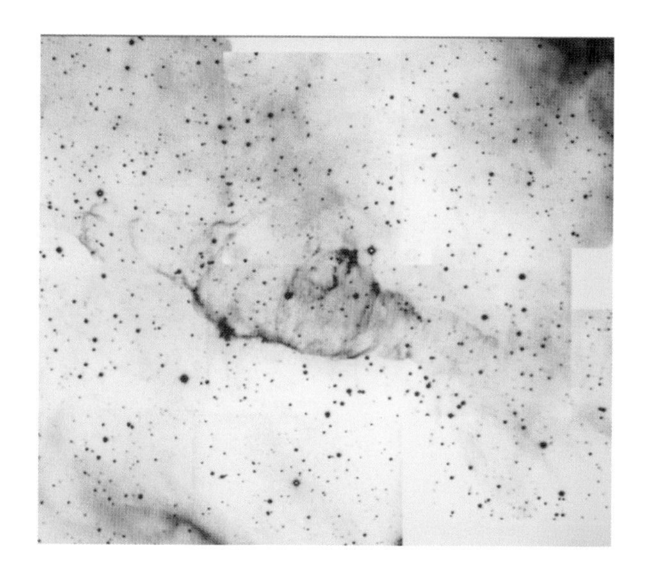

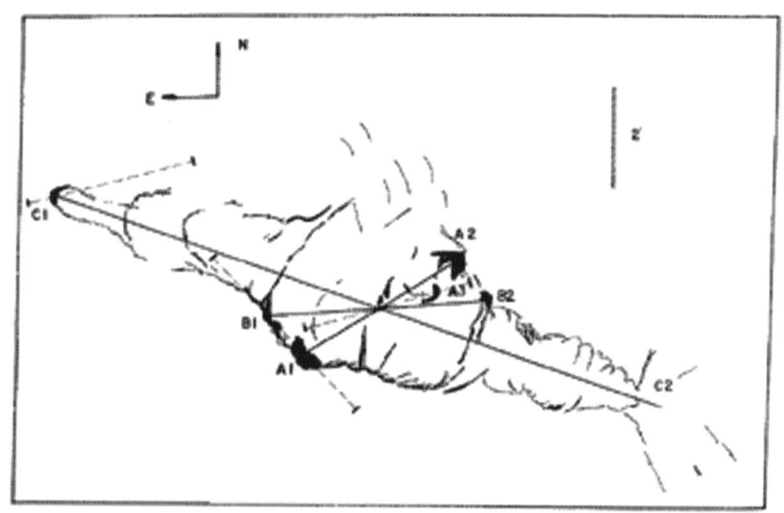

Figure 15.17.
[above] KjPn 8 in Cassiopeia is one of the largest bipolar planetary nebulae known, with lobes extending over 14 arcmin × 4 arcmin in the sky. At a distance of 3300 light years, its size exceeds 1.3 light years. The core (the bright spot at the center of the nebula) was resolved by the *HST* and is found to have the shape of an elliptical ring and a diameter of only 7 arcsec. The three pairs of opposite knots are illustrated in the schematic diagram on the right (credit: J.A. López, R. Vázquez and L.F. Rodríguez).

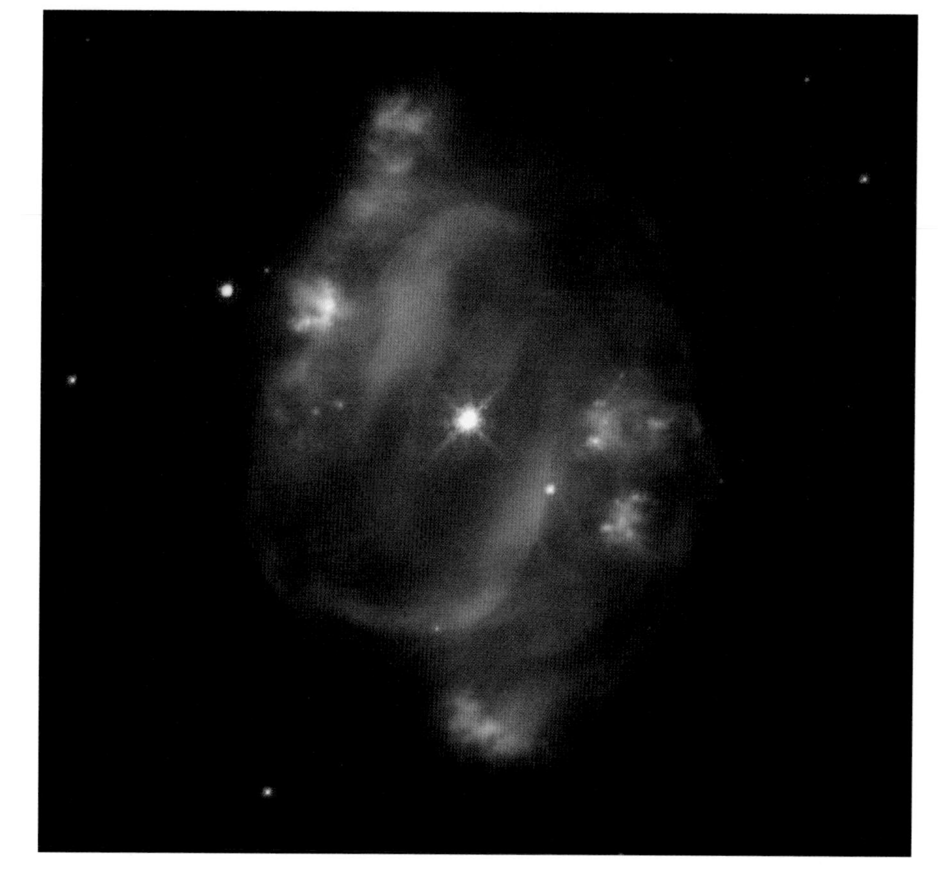

Figure 15.18.
[right] NGC 5307 resembles a pinwheel galaxy more than a conventional view of a planetary nebula. This small (longest dimension about 18 arcsec) planetary nebula represents an example of point-symmetric structure.

Figure 15.19.
[opposite] He3-1475 has a classic point-symmetric structure which counters our preconceptions of what planetary nebulae should look like.

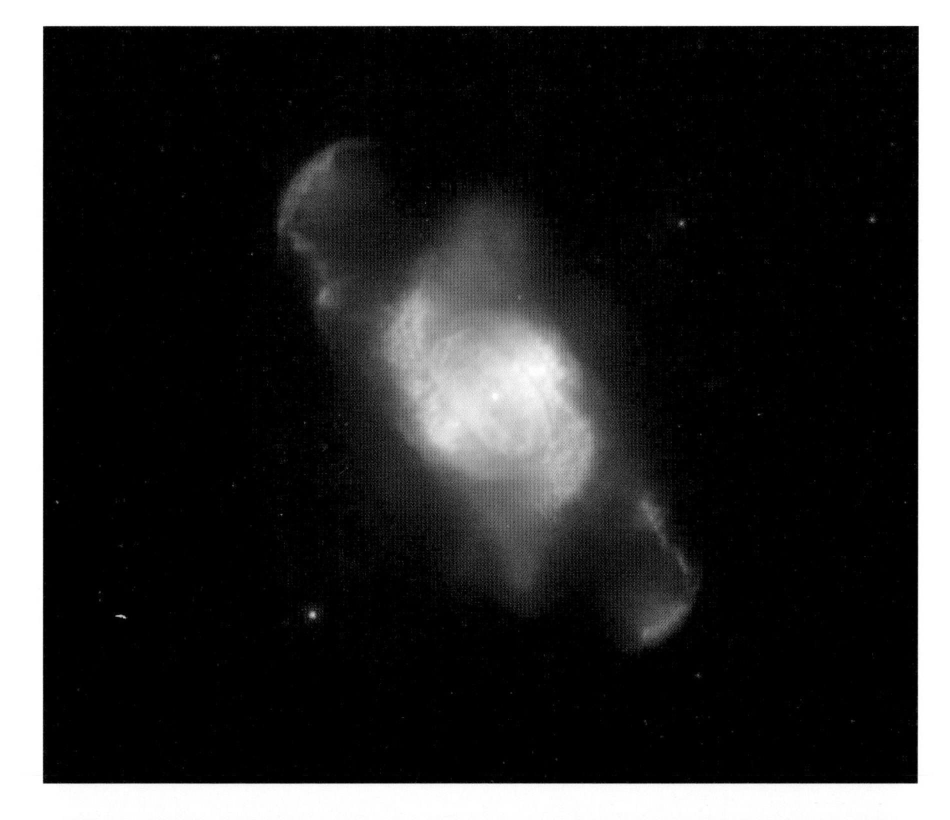

Figure 15.20.
IC 4634 shows a central ring with two pairs of lobes and FLIERs at the end of the major axis.

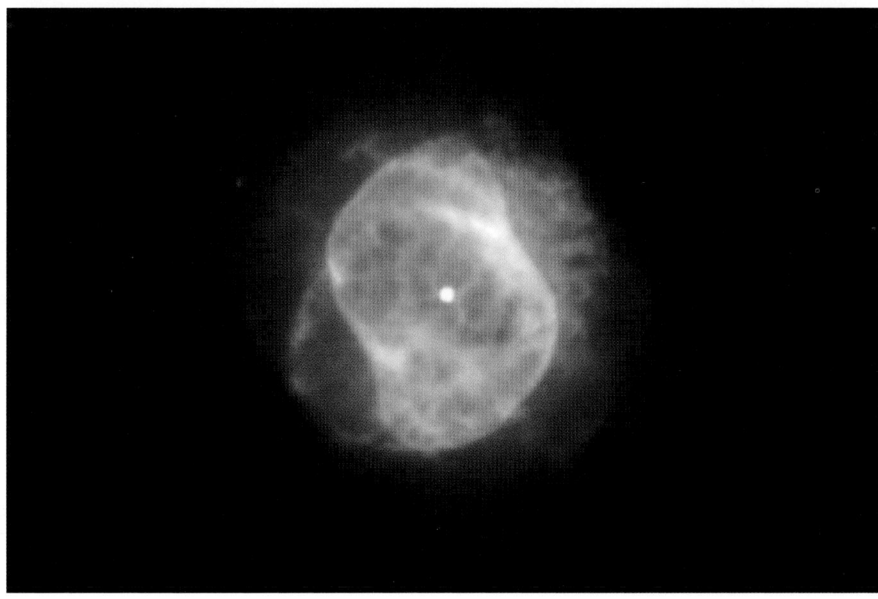

Figure 15.21.
The spiral structure seen in this *HST* WFPC2 image of NGC 6884 has been interpreted by Luis Miranda of the *Instituto de Astrofisica de Andalucia,* Spain, as the result of a pair of precessing jets with a period of about 500 years.

of more than 20 nearly circular arcs can be seen in the *HST* image of the Egg Nebula (Fig. 13.1). Even more perfectly concentric arcs (although fewer in number) are seen in the Cotton Candy Nebula (Fig. 14.1). Since these faint arcs lie on top of the bright nebulae, only the *HST* has the sensitivity to detect them. Soon after, similar circular arcs were seen in the planetary nebulae NGC 6543 (Fig. 15.22), NGC 7027, and Hubble 5. Even more puzzling, the intervals between the arcs are very uniform. We believe that these arcs are projections of spherical shells onto the plane of the sky. If these spherical shells are part of the remnant of the asymptotic giant branch wind, then they seem to suggest that the wind was not steady but came out in 'puffs'. What causes these puffs is a mystery that is yet to be resolved.

Just when we feel that we have a good theoretical grasp of the nature of planetary nebulae, things like arcs, FLIERs, BRETs, arcs and cometary knots make havoc in the community. Perhaps it is Nature's way of saying that she is smarter than we are.

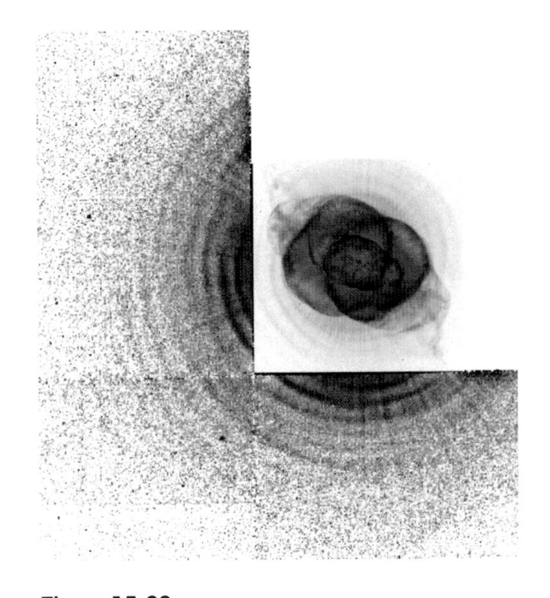

Figure 15.22.
A deep exposure of the Cat's Eye Nebula (NGC 6543) by the *HST* has revealed at least nine circular, concentric rings. In this picture, we have replaced the central overexposed part of the image on the Planetary Camera chip (upper right corner) by an unsaturated image from another observation, showing the relation between the rings and the nebula.

16 How many are there?

Most of the familiar planetary nebulae, like the Ring and Helix, are within a few hundred light years of the Sun. The Milky Way galaxy (Fig. 16.1) is about 100 000 light years across, so the total number of planetary nebulae must be much larger than our local sample (Fig. 16.2). From the observed number density of planetary nebulae in the solar neighborhood, astronomers estimate that there are about 15 000 planetary nebulae total in the Galaxy. Since each planetary nebula lives about 15 000 years, there is approximately one planetary nebula born every year in the Galaxy.

How does this number compare with the actual number of planetary nebulae detected? While nearby planetary nebulae can be identified by their nebular morphology, distant planetary nebulae are small and difficult to find. The last comprehensive planetary nebulae catalogue by Agnès Acker of *Strasbourg Observatory* in 1992 contained just over 1000 planetary nebulae. About 500 more have since been discovered by the *Anglo-Australian Observatory/United Kingdom Schmidt Telescope* survey of the southern galactic plane. Many of these nebulae are so small that they are stellar in appearance, and are known to be planetary nebulae only through their emission line spectra. The factor of ten discrepancy between the predicted 15 000 and the actually catalogued 1500 can be attributed to several factors. One is incompleteness of the survey. There has been no large-scale spectroscopic survey of the entire sky, and many planetary nebulae are probably not catalogued because we never look for them.

Figure 16.1.
[opposite] A panoramic view of the Milky Way. The dark patches are due to absorption by interstellar dust along the line of sight. The center of the Milky Way galaxy is located in the constellation Sagittarius, close to the border of Scorpius and Ophiuchus. This fisheye photograph of the Milky Way was taken in Australia shortly after evening twilight by Wei-Hao Wang.

Another factor is dust extinction in the Galaxy. The Sun is located in the outer parts of the Galaxy, and the dust that lies between us and the Galactic center blocks our view to the other side of the Galaxy. Half of all planetary nebulae could be missed this way. Some planetary nebulae are simply too faint for our current technology to detect. As a nebula ages, its radiation gets weaker and weaker, and distant, old planetary nebulae are difficult to see.

Another possibility is that we may have erred in the estimation of the total number. In order to extrapolate between the local planetary nebulae density to the total, we need to know how far away the nearby planetary nebulae are. For example, if we had underestimated the distances to nearby nebulae, the local density would be lower, and so would the total number. Unfortunately, our ability to measure astronomical distances is rather poor. We are faced with a two-dimensional sky with no sense of depth or distance. For most stars, at least those on the main sequence, we have some theoretical knowledge of the relation between their color and brightness. For example, a star that has similar colors to the Sun should have a similar luminosity. By comparing the apparent brightness of the star and its supposed intrinsic brightness, we can derive its distance. However, for planetary nebulae there is no such clear cut color–luminosity relationship, and two nebulae can have vastly different luminosities depending on the mass of the star they descended from.

In 1956, Shkolvsky pioneered a technique to derive distances from planetary nebulae by assuming that they all have the same mass in the nebulae. This was a reasonable assumption at the time because he thought that planetary nebulae were just detached envelopes of red giant stars. However, in our modern interacting winds picture, the nebular mass will increase with time as more and more mass is shoved up by the snow plow! By 2000, astronomers had abandoned the Shkolvsky method but failed to find an alternative reliable method to determine distances. Owing

to this uncertainty in distances, we could have a factor of 2 error in the total number of planetary nebulae in the Galaxy.

A good estimate of the total number of planetary nebulae is important for other reasons. In Chapter 6, we learned that 95% of all stars will become planetary nebulae. If we know the exact birth rate of planetary nebulae (one per year is our best guess at this

Figure 16.2.
[opposite] A map of the Milky Way galaxy showing the positions of the Sun and some of the nearby planetary nebulae. Our solar system is located in the outer regions of the Milky Way galaxy, about 28 000 light years from the galactic center and about 20 light years above the galactic plane. The Sun (and the solar system) is orbiting around the galactic center on a nearly circular orbit with a speed of about 250 km/s. Since the formation of the solar system about four and half billion years ago, we have gone around the galaxy about 20 times.
The Milky Way is believed to contain at least 200 billion stars, with a total mass on the order of one trillion solar masses. The Milky Way is the second largest galaxy (after M31, the Andromeda Galaxy) in the local group. Our closest neighboring galaxy is the Sagittarius dwarf elliptical galaxy, located about 50 000 light years from the galactic center. A little further away are the Large and Small Magellanic Clouds at 162 000 and 200 000 light years respectively.

moment), then we know the death rate of stars. Since most of the planetary nebulae descended from low-mass stars that were born billions of years ago, we therefore know the stellar birth rate in the past! This has significant implications on our understanding of the early evolution of our own Galaxy.

One of the ways to check the various distance determination methods is to observe planetary nebulae in stellar systems with well-known distances. One such possibility is to use globular clusters. Globular clusters are collection of stars held together by mutual gravitational attraction. They often contain thousands, or even millions of stars that were born at the same time together. In contrast to the galactic clusters discussed in Chapter 9, stars in globular clusters were formed a long time ago during the early years of the Galaxy. The average age of stars in globular clusters is about 12 billion years old. Since massive stars evolve much faster than low mass stars, all the massive stars in globular clusters have died long ago, and all remaining stars are of low mass.

In Fig. 16.3 we show a picture of M15, one of the brightest globular clusters in the Milky Way galaxy. On August 30, 1927, F.G. Pease discovered a planetary nebula (since named Pease 1) in the cluster, the first planetary nebula to be found in a globular cluster. Since the distance to M15 (37 000 light years) is well known, the physical properties (such as mass and size) of this planetary nebula can be determined more accurately than those of other galactic planetary nebulae. Unfortunately, among the approximately 200 globular clusters present in our Galaxy, only four are known to contain planetary nebulae (one in each). The rarity of planetary nebulae in globular clusters makes them not very useful as distance standards. Astronomers are therefore forced to look further, beyond our Galaxy.

Our galactic neighbors, the Large and Small Magellanic Clouds, are known to harbor at least 400 planetary nebulae. Because of their emission line spectra, astronomers are able to distinguish the

Figure 16.3.
M15 in Pegasus was discovered by Jean-Dominique Maraldi on September 7, 1746, while he was looking for a comet. He described it as 'a nebulous star, fairly bright and composed of many stars'. M15 is one of the 29 globular clusters listed in Messier's catalogue and one of nearly 150 globular clusters in the Milky Way Galaxy. M15 is also the first globular cluster found to contain a planetary nebula. With the 100-inch telescope at *Mt. Wilson*, Pease found a planetary nebula just outside the cluster core. In this *HST* picture imaged with narrow-band filters, the planetary nebula Ps 1 clearly stands out in the upper left corner of the crowded stellar field of the cluster (credit: NASA and the Hubble Heritage Team, Space Telescope Science Institute).

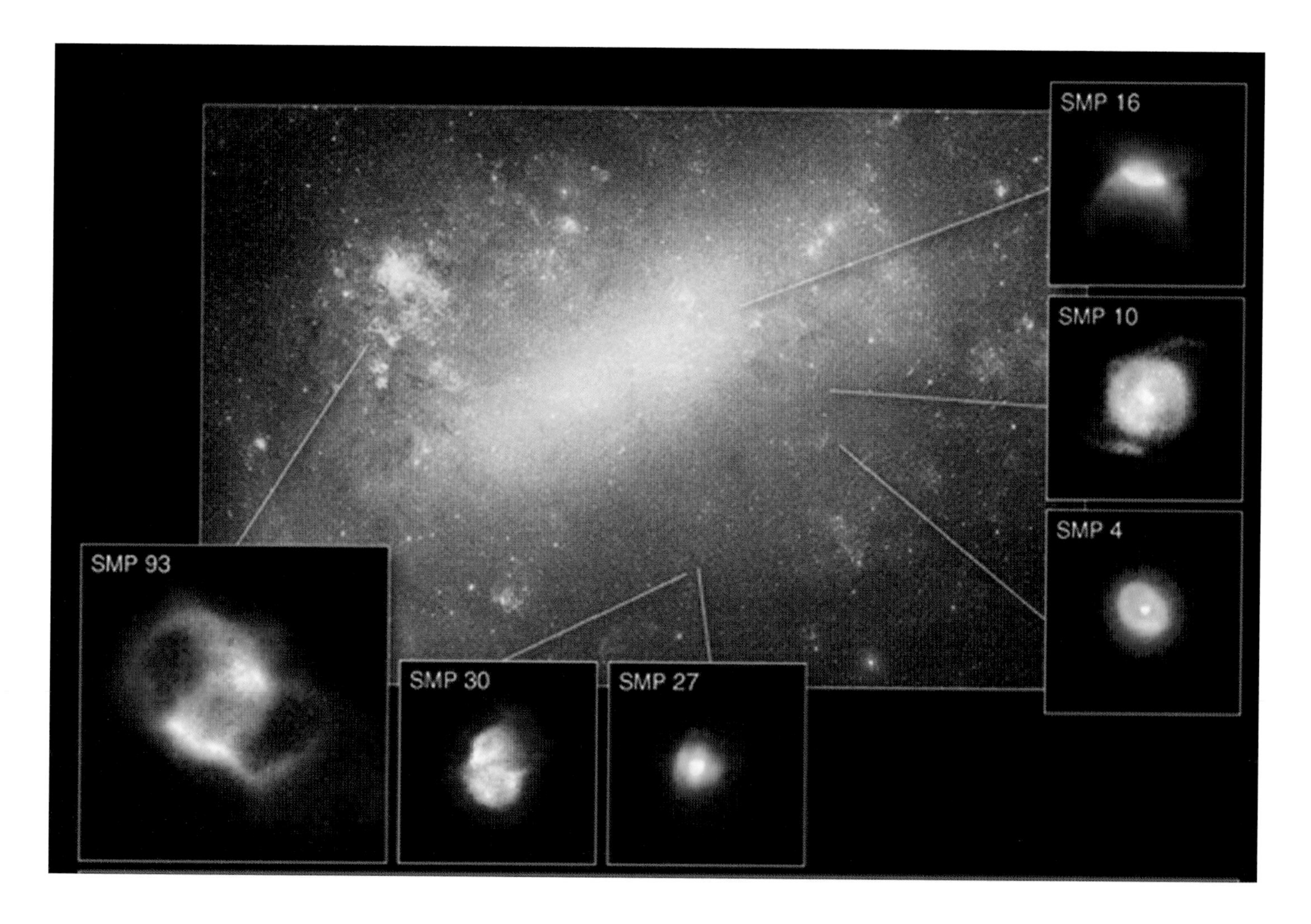

Figure 16.4.
Six planetary nebulae in the Large Magellanic Cloud imaged by the *Space Telescope Imaging Spectrograph* (*STIS*). The lines point to the objects' location in the Large Magellanic Cloud. Three of the nebulae (SMP 16, 30, and 93) have bipolar shapes, whereas SMP 10 has a pinewheel shape similar to that of NGC 6543 (credit: L. Stanghellini). The ground-based image of the Large Magellanic Cloud is by D. Malin.

nebulae from stars in spite of the crowded star field in these two galaxies. With the superior angular resolution of the *HST*, some of these nebulae can even be imaged. Six examples of Large Magellanic Cloud planetary nebulae are shown in Fig. 16.4. We can see that they have elliptical and bipolar forms just the same as galactic planetary nebulae.

The existence of planetary nebulae is not limited to our own Galaxy or our immediate neighbors. Because planetary nebulae have very different spectroscopic signatures from stars, they can be identified even in galaxies far away. By placing a narrow-band filter in front of telescopes and preferentially selecting light from the emission line of oxygen, astronomers can easily pick out planetary nebulae from a crowded field of stars in a galaxy. Walter Baade was the first to have discovered planetary nebulae in M31, the Andromeda galaxy, in 1955. In 1978, Holland Ford and George Jacoby detected over 300 planetary nebulae in M31. With 4-m class telescopes, planetary nebulae can be seen as far as the Virgo and Fornax clusters of galaxies. It is gratifying to know that stars in other galaxies evolve in ways similar to those in our own Galaxy.

17 Measuring the size and mass of the Universe with planetary nebulae

Over the past 60 years, the size of the Universe has greatly increased, not because the Universe has expanded but because of the ever-decreasing value of the Hubble constant adopted by astronomers. The Hubble constant, which determines the age of the Universe, is estimated using the distance to external galaxies and the observed receding rate (red shifts) of the galaxies. Since astronomers' guesses of extragalactic distances have been extremely poor, the value of the Hubble constant has changed by a factor of ten since it was first measured by Edwin Hubble. Even as recently as a few years ago, there were protagonists who held views on the value of this important parameter that differed by a factor of 2.

The solution to this problem is very simple in principle. We need to find a 'standard candle' that is bright enough to be seen at large distances, yet has a narrow and uniform intrinsic brightness. By comparing the apparent and intrinsic brightness of these standard candles, we can derive how far away they are. Various objects have been used, including Cepheid variables, novae, supernovae, etc., with varying degrees of success.

The use of planetary nebulae as a 'standard candle' was pioneered by George Jacoby of *Kitt Peak National Observatory*. By putting a filter in front of the telescope and therefore allowing only the light of oxygen atoms into the telescope, planetary nebulae could easily be isolated and distinguished from stars even in very distant galaxies (Fig. 17.1).

Figure 17.1.
[opposite] By imaging in a narrow-band mode centering on the atomic oxygen line, planetary nebulae are easily separated from stars and are commonly detected in galaxies outside of the Milky Way. This image identifies the planetary nebulae (shown as circles) in Ursa Major's grand spiral galaxy M101. By using these nebulae as standard candles, George Jacoby found a distance to that galaxy of 25.1 million light years (credit: John J. Feldmeier, Robin Ciardullo, and George Jacoby).

Using the thousands of planetary nebulae in nearby galaxies as a sample, Jacoby and his collaborators found that the brightest planetary nebulae in a galaxy always have the same intrinsic brightness, even in galaxies of very different types. When he applied this technique to the determination of the Hubble constant, he derived a value which was much larger than what was then fashionable. These new results implied that the Universe is only ten billion years old. Recent *HST* observations using the traditional technique of Cepheid variables have lent support to Jacoby's results.

One of the most exciting astronomical discoveries in the past 30 years has been our realization that the Universe is made up of more than the stars and galaxies. Most of the mass of the Universe, astronomers now believe, is made up of mysterious, unknown forms of matter known as 'dark matter'. It is labeled as such because dark matter does not shine, and its existence is only inferred by its gravitational effects on visible objects. In the Milky Way galaxy, most of the stars lie on the plane of the Galaxy which astronomers call the 'disk'. The Sun, for example, lies on the disk about 25 000 light years from the center of the Milky Way. Stars can also be seen in the halo, the spherical volume that surrounds the disk. Although there are fewer stars in the halo than in the disk, most of the mass of the Milky Way is in the halo, in the form of dark matter.

Since planetary nebulae exist not only in a galaxy's plane but also in the halo, their orbital motions are affected by the gravity from normal luminous matter (stars and interstellar gas), as well as by the large amount of invisible dark matter in the halo. With current 4-m class telescopes, the velocities of planetary nebulae can be measured with relatively high precision, and the distribution of a galaxy's mass – both luminous and dark – can be traced. Painstaking velocity measurements on planetary nebulae have also allowed astronomers to 'weigh' galaxies and their dark-matter halos.

Figure 17.2.
[opposite] The elliptical galaxy NGC 5128 is also known as the Centaurus A for being the brightest radio source in the constellation of Centaurus. This galaxy is one of the most luminous and massive galaxies known and is a strong source in X-rays, gamma rays, and the infrared, as well as in the radio. The dark band across the galactic disk could be the result of giant explosions in the nucleus, or debris from a smaller dusty galaxy that is being absorbed by Centaurus A (credit: *Cerro Tololo InterAmerican Observatory* 4-meter Blanco Telescope/AURA/NOAO/NSF).

The first work to use planetary nebulae as tracers of dark matter was done by Xiaohui Hui in her Ph.D. thesis research at Boston University. Using the prime focus CCD camera of the 4-m telescope at the *Cerro Tololo InterAmerican Observatory* (*CTIO*), she identified 785 planetary nebulae in the galaxy NGC 5128. NGC 5128, also known as Centaurus A (Fig. 17.2), was one of the first examples of two galaxies in collision found by Baade and Minkowski. This was followed by spectroscopic observations at the *Anglo-Australian Telescope* and *CTIO*. An instrument called a multifiber spectrograph allowed light to be collected from many planetary nebulae and fed by optical fiber to have their spectra recorded simultaneously. By measuring the wavelengths of the doubly ionized oxygen line and using the principle of Doppler Effect, she was able to derive the velocities of 433 planetary nebulae in the halo of Centaurus A to an accuracy of 4 km/s. Since the motions of planetary nebulae are affected by the distribution of mass in the galaxy, an analysis of the planetary nebulae velocities showed that the galaxy contains much more mass than implied by the visible light. There is definite evidence for the presence of dark matter in the halo of the galaxy. With 8-m and 10-m telescopes in use today, this method can be extended to galaxies as far as the Virgo cluster.

In most of this book, we have been concerned with the origin and structure of planetary nebulae. There is now, however, increasing attention being paid by astronomers to employing planetary nebulae as a tool to study our galactic environment. This includes the use of planetary nebulae as distance indicators or as tracers of dark matter, as we discussed above. Since planetary nebulae also represent that end stage of a star's evolution, its chemical makeup, as determined by spectroscopic observations, can also tell us about the nuclear synthesis history of its progenitor star. By observing planetary nebulae in different galaxies, or even between the disk and halo of our own Galaxy, we can learn about whether stars in different environments evolve differently. This is akin to a physician examining a group of dying persons, and trying to diagnose whether they smoked or drank alcohol during their youth, and whether such habits can be correlated to where they grew up. I call this practice 'stellar sociology'.

18 Old stars as molecular factories

Although planetary nebulae represent the end of a star's life, their existence may have also sown the seeds of new stars, new planets, and even life itself. Carbon and nitrogen, two of the elements that form the basis of life on Earth, are made in asymptotic giant branch stars. The cells in our body are mostly made up of organic molecules composed of hydrogen, nitrogen, oxygen, phosphorus, and calcium atoms in carbon-based structures. Every one of the carbon atoms in our body was once created in an asymptotic giant branch star. In the early 1950s, astronomers knew that carbon atoms could be synthesized from helium atoms by nuclear reactions. However, calculations showed that any such carbon atoms formed will quickly turn into oxygen, and very little carbon will be left. This is in contradiction to the observed fact that carbon is the fourth most abundant element in the Universe (after hydrogen, helium, and oxygen) and its wide presence is required for the creation of life on Earth. In order to solve this problem, Fred Hoyle of the University of Cambridge in 1953 predicted a hypothetical nuclear state of carbon. If such a state exists, carbon will not entirely be converted to oxygen. Hoyle's prediction was quickly confirmed experimentally by Willy Flower of California Institute of Technology, resulting in a Nobel Prize for Flower in 1983.

Armed with the knowledge of nuclear physics, astronomers found that many heavy elements, including Y, Zr, Ba, La, Ce, Pr, Nd, Sm, Eu, etc. can be made in asymptotic giant branch stars. With the emergence of computers in the 1960s, astronomers began to incorporate detailed nuclear reactions into stellar structure calculations. Numerical models by Icko Iben of the University of Illinois and others showed that carbon atoms synthesized near the core of asymptotic giant branch stars can be 'dredged up' to the surface during brief episodes of helium burning. Since the surfaces of asymptotic giant branch stars are cool, molecules can form in the stellar atmosphere. Late in the life of asymptotic giant branch stars, carbon-based molecules such as CO, CN, C_2, and C_3 can be detected by spectroscopic observations. The stars that show these molecules are known as carbon stars.

There are several thousand known carbon stars in the Galaxy. The most famous member is CW Leo, which we have already mentioned in Chapter 8. This star is losing mass at a high rate, and is believed to be within a few thousand years of evolving off the asymptotic giant branch to become a proto-planetary nebula. Using increasingly sensitive radio telescopes, astronomers have detected close to 50 molecules in the stellar wind of this star. The list of molecules detected includes ammonia (NH_3), HC_2NC, HNC_3, CH_3CN, acetylenic radicals such as C_2H, C_3H, ... to C_8H, and long-chain linear molecules (called cyanopolyynes) such as HCN, HC_3N, ... to HC_9N, etc. Carbon stars are amazing molecular factories.

Since planetary nebulae are formed from the remnants of circumstellar envelopes of asymptotic giant branch stars, such molecules may still be present in planetary nebulae. In fact, the detection of carbon monoxide in the planetary nebula NGC 7027 was one of the first cited pieces of evidence in support of the interacting winds theory (see Chapter 10). In spite of the hostile environment, where the temperature is high and the ultraviolet radiation field is strong, many molecules have been discovered in planetary nebulae.

Between 1970 and 2000, most of the interstellar molecules were found by millimeter-wave telescopes detecting radiation generated by molecules as they change from one rotational state to another.

However, this technique is less useful in the search for complex molecules. The rotational spectra of many organic molecules are weak and difficult to find with millimeter-wave techniques. Fortunately, many organic molecules have strong vibration and bending modes, which radiate in the infrared parts of the electromagnetic spectrum. With the advancement of infrared observing capabilities, the search for large, complex molecules became a possibility.

Owing to absorption by water vapor, the Earth's atmosphere is opaque to most infrared radiation. Ground-based observations, like those we described in Chapter 8, can peek through certain windows where the water absorption is less severe. Even at high altitude sites such as Mauna Kea, observing is impossible over much of the infrared wavelengths. The ideal place to perform infrared spectroscopic observations is therefore in space.

Although the *IRAS* satellite was most well known for the infrared images it took of the Galaxy, it also carried an infrared spectrometer built by Thijs de Graauw of the Laboratory for Space Research in the Netherlands. As the *IRAS* satellite surveyed the entire sky, the spectrometer, called the Low Resolution Spectrometer (LRS), also took infrared spectra of every infrared source that was bright enough to allow a spectrum to be recorded. At the end of the mission, spectra of over 50 000 stars were obtained, providing a first glimpse of the spectral properties of stars in the mid-infrared parts of the spectrum.

Many of the objects detected by the *IRAS* LRS were asymptotic giant branch stars, both the oxygen- and carbon-rich varieties. Next were young stars that are newly born and still surrounded by the molecular clouds from which they were formed. Although a number of galaxies were detected, the sensitivity of the LRS was inadequate to study these far away objects.

The fact that the LRS was an unbiased survey of old stars in the Galaxy represented an unprecedented opportunity to learn about the global behavior of old stars. This survey allows us to give an accurate determination of how many of the asymptotic giant branch stars in the Galaxy are oxygen rich and how many are carbon rich, and therefore gives us the clues on how they form. Paul Wesselius of the Laboratory for Space Research kindly gave me and my colleagues access to the entire LRS database, and we proceeded to examine this treasure at the University of Calgary. With great care, Kevin Volk extracted and analyzed each spectrum, and classified them into ten astrophysical groups based on the spectral behavior. Bill Bidelman of Case Western Reserve University, a world expert on the spectra of stars, collaborated with us in correlating the infrared spectra with his extensive collection of optical spectra. By the time our funding ran out, we had produced a comprehensive infrared spectral catalogue containing over 11 000 stars.

While we had no problem assigning most of the spectra into well-known classes, there were a number of peculiar spectra that were completely different. Kevin Volk gave them the assignment of 'U', standing for 'unknown'. In particular, several of them showed a previously unseen emission feature at the wavelength of 21 μm. This feature was so strong that it could not have been an instrumental effect. With the *United Kingdom Infrared Telescope*, we identified the optical counterparts of these sources, and I immediately recognized that these stars looked exactly the same as the proto-planetary nebulae that Bruce Hrivnak and I have been studying (Fig. 18.1). In other words, this mysterious emission feature is produced during the proto-planetary nebulae phase of evolution (see Chapter 13).

In December 1988, I presented these results at the European Space Agency meeting on infrared spectroscopy at Salamanca, Spain. We reported four proto-planetary nebulae, all previously unknown *IRAS* sources, that show this unidentified emission feature. Using the *United Kingdom Infrared Telescope* and working with Tom Geballe of the Joint Astronomy Center in Hilo, Hawaii, we proceeded to look for new sources showing the 21-μm emission.

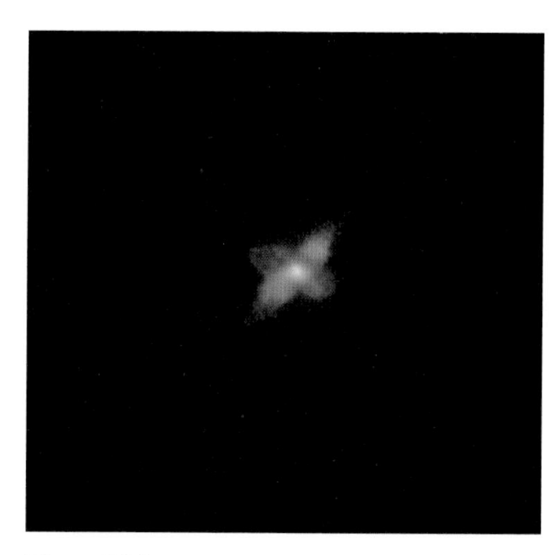

Figure 18.1.
HST WFPC2 image of *IRAS* 04296+3429, one of the first 21-μm sources discovered by Kwok, Volk, and Hrivnak in 1989. Although it was identified as a proto-planetary nebula at the time, it was not until ten years later that the structure of the object was revealed by the *HST*.

Ten years later, when I reviewed this subject at the International Astronomical Union symposium on asymptotic giant branch stars in Montpellier, France, we had discovered 12 such objects.

Astronomers were totally puzzled by this discovery. Since the feature is strong, it has to be primarily made up of common elements such as hydrogen, oxygen, or carbon. By that time, we had already determined by ground-based optical spectroscopy that these stars are carbon-rich, logically we suspected that the carrier of this feature is made up primarily of carbon. But what form of carbon? Two forms of carbon – graphite and diamond – have long been known to exist on Earth. Graphite is the soft, black substance that we use for pencils; diamond is the hard, colorless stuff that we acquire at high cost to put around our fingers. However, the infrared spectra of neither graphite nor diamond show any resemblance to the 21-μm sources.

In 1995, Robert Curl, Harry Kroto and Richard Smalley, in an effort to find the mechanism of forming long-chain molecules in the circumstellar envelopes of carbon stars, discovered a new form of carbon. They showed that 60 carbon atoms can be arranged in the form of a soccer ball-like cage and remain stable. This structure, with 32 faces consisting of 12 pentagons and 20 hexagons, was named buckminsterfullerene, or buckyballs for short. For this work, Curl, Kroto and Smalley were awarded the Nobel Prize in chemistry in 1996.

Could the 21-μm feature be due to this new form of carbon? This possibility was investigated by Andrew Webster of the University of Edinburgh. By performing theoretical calculations on the structures of the buckminsterfullerenes, he found that C_{60} molecules with different numbers of hydrogen atoms attached to the corners could produce an emission feature similar to that observed by us. Other adventurous ideas were also proposed by several other groups, including the suggestion of very small diamonds by H.G.M. Hill and Louis d'Hendecourt of the University of Paris.

Because of water absorption in the Earth's atmosphere, there are limits on the quality of infrared observations that can be carried out on a ground-based telescope. The launch of the *ISO* satellite in 1995 (see Chapter 10) gave us the opportunity to study the 21-µm feature in much greater detail. By comparing the *ISO* spectra of several 21-µm sources, we measured the peak wavelength of the feature to be at 20.1 µm. We were also able to determine an accurate shape of the emission feature and concluded that the carrier cannot be a small molecule, but a large molecular cluster or a solid.

Totally unaware of these astronomical developments, Gert von Helden of Matter's Institute for Plasma Physics in Nieuwegein was leading a team to synthesize titanium carbide (TiC) in the laboratory. They were able to make small clusters of TiC with 27–125 atoms. They then bombarded these small molecular units (called nanoclusters) with lasers. When the laser wavelength resonates with a TiC emission feature, the laser's photons cause the cluster to emit an electron. With this technique, they showed that TiC has a strong emission feature at 20.1 µm. This agrees precisely with the peak wavelength that we measured from our *ISO* observations.

The element titanium (named after the giant Titans in Greek mythology) is well known for is strength and light weight and is frequently used in aircrafts and missiles. The molecule titanium oxide (TiO) is also commonly observed in the atmospheres of oxygen-rich asymptotic giant branch stars. So the suggestion of TiC as the carrier of the 21-µm feature is not unreasonable. There are, however, questions on whether there is a sufficient amount of the titanium element in these stars to explain the strong 21-µm feature observed. If this identification turns out to be correct, then proto-planetary nebulae must be remarkably efficient molecular factories in manufacturing these molecular clusters.

19 Do we owe our lives to planetary nebulae?

In 1977, a team of astronomers at University of California at San Diego consisting of Ray Russell, Tom Soifer and Steve Wilner used NASA's *Kuiper Airborne Observatory* (*KAO*) to observe the planetary nebula NGC 7027. The *KAO* (Fig. 19.1) was converted from a C-141 military transport aircraft with a side opening to allow for the installation of a 0.9-m (36-inch) telescope. Flying at 13 000 m, the *KAO* can avoid more than 99% of the water vapor in the Earth's atmosphere. To the astronomers' surprise, they discovered a series of infrared emission features that had never been seen before. They called these features Unidentified Infrared (UIR) features. They thought the nature of these features would soon be known, but as it turned out, this name stuck for almost 20 years.

The first step toward the identification of the UIR features came in 1979. Walt Duley of York University in Canada and David Williams of the Manchester Institute of Science and Technology suggested that these features are due to the stretching and bending modes of aromatic molecules. An aromatic molecule has a ring structure made up of six carbon atoms, of which benzene (C_6H_6) is the simplest example. This cause was taken up by a number of groups, including the NASA Ames Research Center astrochemistry group led by Lou Allamandola. From laboratory studies, they suggested that a class of molecules called polycyclic aromatic hydrocarbons (PAH) is responsible for the UIR features. PAH molecules are benzene rings bonded together in a planar structure. An example is naphthalene ($C_{10}H_8$), which is commonly used as a

Figure 19.1.
[opposite] The control room of the *Kuiper Airborne Observatory* showing the control panel for the telescope along the right side of the aircraft. During observing flights, four Ames crewmembers control the operation of the telescope. The television monitors are used in flight to display navigation maps, system status and video pictures of stars from the telescope's three cameras (credit: NASA Ames Research Center).

moth repellent. Another example is benzopyrene ($C_{20}H_{12}$), a cancer causing (carcinogenic) substance found in cigarette smoke, chimney soot, and barbecued meat. PAH molecules are most often encountered in our everyday life as the result of incomplete combustion. For example, while most of the hydrogen atoms in gasoline are burned (oxidized) into water, not all of the carbon atoms are converted into carbon dioxide (CO_2). The remaining carbon-rich substance is the black stuff coming out of the exhaust pipe.

However, these stretching and bending modes are not unique to PAHs, and other aromatic molecules can also give similar infrared

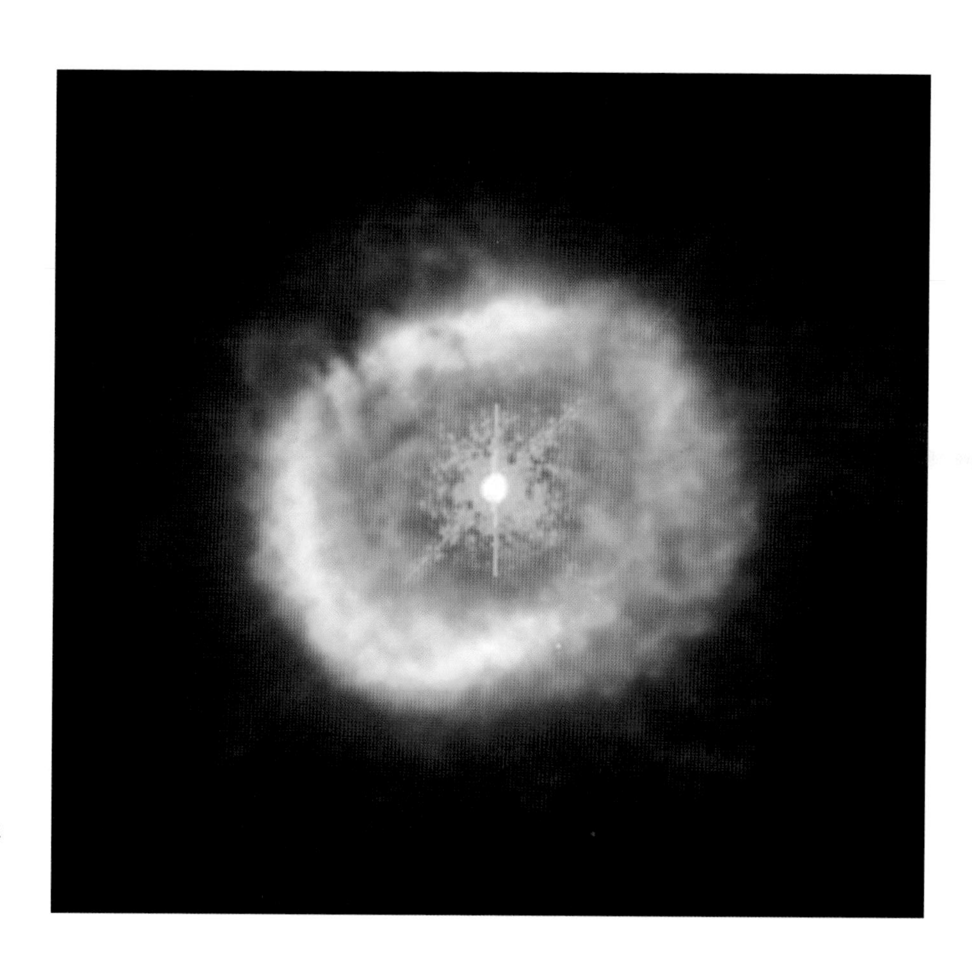

Figure 19.2.
BD+30°3639, a small (arcsecond) planetary nebula in the constellation of Cygnus, has a carbon-rich chemical composition similar to that of NGC 7027. *ISO* observations have found that this nebula is rich in aromatic molecules.

signatures. Walt Duley, using lasers to blast graphite, created a synthetic material called hydrogenated amorphous carbon (HAC) which, when examined, shows the same infrared features as observed in NGC 7027.

Regardless of the exact nature of the stuff that gives rise to the UIR features, it is clear that it is a complex organic compound with ring-like structures similar to many of the same organic molecules that are in the bodies of living organisms. This suggests that stars not only make atoms and simple molecules, but also complex organic molecules (Figs. 19.2 and 19.3)!

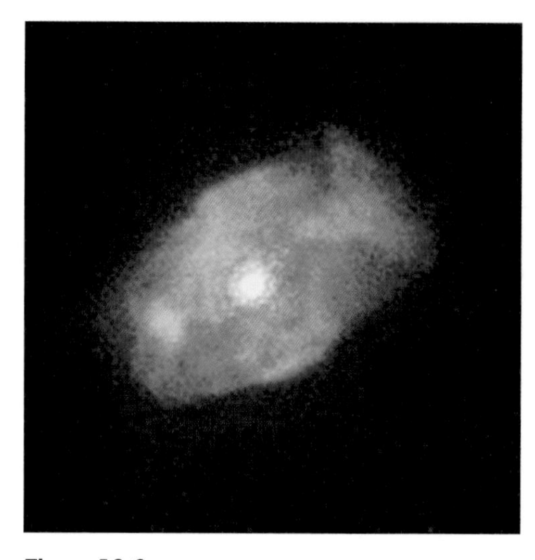

Figure 19.3.
Another planetary nebula rich in organic molecules is *IRAS* 21282+5050. This small nebula is very bright in the infrared and was discovered by the *IRAS* satellite.

Intrigued by the discovery of such aromatic compounds in planetary nebulae and having discovered a number of proto-planetary nebulae ourselves, we were interested in investigating the infrared spectra of proto-planetary nebulae for early signs of the presence of aromatic compounds. We were particularly interested in observing carbon-rich proto-planetary nebulae. Without oxygen to tie up the carbon atoms into uninteresting molecules such as CO, there are many more possibilities for synthesizing a variety of organic compounds. With an abundant supply of hydrogen, carbon and hydrogen can combine into either linear or ring-like structures. We can tell proto-planetary nebulae are oxygen- or carbon-rich because oxygen-rich nebulae often show an emission feature due to an amorphous form of silicates (see Chapter 8). Any nebula without the silicate feature would be an ideal laboratory for carbon-based chemical synthesis.

The opportunity came with the launch of European Space Agency's *Infrared Space Observatory* in 1995. Using the Short Wavelength Spectrometer built by Thijs de Graauw of the Laboratory for Space Research in the Netherlands, we found that the aromatic features seen in planetary nebulae are not as strong in proto-planetary nebulae. Instead, new features due to aliphatic bonds were seen. Aliphatic bonds have a tetrahedral geometry and are different from the planar structure of the aromatic rings. On the other hand, extremely evolved carbon stars (infrared stars with no visible counterparts) show the presence of acetylene (C_2H_2) molecules, which are widely believed to be the building blocks of benzene. Cousins of acetylene, diacetylene (C_4H_2) and triacetylene (C_6H_2), were also found in proto-planetary nebulae by José Cernicharo of the Consejo Superior de Investigaciones Científicas in Spain. The different infrared spectra of carbon stars, proto-planetary nebulae, and planetary nebulae led us to believe that molecules are continuously being synthesized into increasingly complex forms in the circumstellar environment. Since we know the transition from the asymptotic giant branch to planetary nebulae is only a few thousand years, these molecular factories are extremely efficient.

We have to remember that the gas density in these nebulae is much lower than the terrestrial atmospheric density (see Chapter 3). At the present time, we have no idea how chemical reactions can proceed so rapidly in such unfavorable circumstances. Nature can work miracles.

These discoveries have serious implications. If complex organic molecules are made in planetary nebulae, they will eventually be spread into the interstellar medium from which new stars are formed. It is therefore possible that when the solar system was created four and half billion years ago, the ingredients in the solar nebula might have included organic compounds that were essential for the beginning of life.

How did life begin on Earth? Most scientists believed that life began spontaneously with simple inorganic molecules. In 1953, an experiment done at the University of Chicago by Harold C. Urey and his graduate student Stanley L. Miller showed that many organic molecules can be formed by sending electric discharge into a glass flask containing water, methane, hydrogen, and ammonia. Such molecules, if collected in pools and ponds, could form proteins and nucleic acids which later develop into self-replicating molecular structures. Fossil evidence suggests that the first sign of life on Earth, probably in the form of bacteria, was present as early as three and half billion years ago. If complex organic molecules were already available on the early Earth, they could have jump-started the process and made the origin of life on Earth much easier.

Let us contemplate for a moment the significance of planetary nebulae. They are the remnants of dying stars that have a very short, albeit glorious, life. They end their existence by spreading newly synthesized atoms, molecules and dust throughout the Galaxy. After wandering in the interstellar medium for millions of years, some of these ingredients might have aggregated with ejecta from other planetary nebulae to form a dense cloud from which new stars were born. The debris left over from star formation makes comets, asteroids, and planets. Some of the original material in planetary nebulae may have survived and been deposited on a planet from which life emerged. Next time you look through your telescope in your backyard, please take a look at the Ring Nebula. In spite of the large distance separating it from us, it may be closer to our existence than you thought.

20 Glossary

angular resolution
The ability of a telescope to image the fine details of an astronomical object. Generally the larger the telescope and the shorter the wavelength, the higher the possible angular resolution. However, angular resolution of ground-based optical telescopes is usually limited by the atmosphere.

arcsec
Unit of measurement of size in the plane of the sky. A circle is divided into 360 °, each divided into 60 arcmin, and each divided into 60 arcsec. An object 1 arcsec in size is equivalent to viewing a dime at a distance of 20 km away.

asymptotic giant branch
The stage of stellar evolution characterized by the alternate burning of hydrogen and helium in a shell above a carbon–oxygen core.

atom
The basic constituent of matter. The atom is made up of a number of electrons surrounding a nucleus containing protons and neutrons. The most common element in the Universe, the hydrogen atom, has one proton in the nucleus and one electron outside the nucleus.

black hole
A possible end state of the evolution of massive stars. When the remnant of a star collapses to a small volume, its surface gravity could be so high that even light cannot escape from the star. Since a black hole emits no light, its existence can only be inferred from its gravitational effect on neighboring stars.

carbon stars
A class of asymptotic giant branch stars where the abundance of carbon exceeds that of oxygen.

circumstellar envelope
The gas and dust ejected by the stellar wind of asymptotic giant branch stars form a low density envelope surrounding the star. This envelope is created over a period of several hundred thousand years and can be as large as a light year in size. The existence of the circumstellar envelope is detected by infrared and millimeter-wave observations.

core, stellar
Stars can be simplistically described as having a high-density core and a low-density envelope. Nuclear reactions occur only in or near the core. The central stars of planetary nebulae are basically cores of their asymptotic giant branch progenitors and are mostly made up of carbon and oxygen.

dark matter
Unknown forms of matter in the halo of galaxies. Although not detectable in visible light, its existence is inferred from its gravitational effects.

diamond
A mineral made up of carbon atoms arranged in a tetrahedral structure. There is increasing evidence that diamonds are present in the interstellar medium as a consequence of chemical synthesis in the circumstellar environment.

Doppler Effect
The change of light color as the result of movement of the emitting object. An object moving towards or away from us will have its light shifted to the blue or red respectively.

dredge up
The movement of elements synthesized by nuclear reactions from the core to the surface of an asymptotic giant branch star.

dust
Solid particles of approximately micrometer size. Dust in the galaxy can be detected through its absorption of background starlight and its self emission in the infrared.

electrons
Negatively charged particles surrounding a nucleus of an atom.

envelope, stellar
The outer parts of a star (*see* core, stellar) and are mostly made up of hydrogen. Because of mass loss during the asymptotic giant branch, most of the stellar envelope is lost through a stellar wind. This results in the exposure of the bare core, which forms the central stars of planetary nebulae.

Earth's atmosphere
The Earth's atmosphere represents the greatest obstacle for astronomical observations. It not only distorts the visible starlight (makes them twinkle), but the molecules (in particular water and oxygen) in the atmosphere absorb most of the light from the Universe. Only visible, radio, and some of the infrared light can penetrate the atmosphere and be detected by ground-based telescopes.

FLIERs
Fast, low-ionization, emission regions usually found in pairs along the axis of planetary nebulae.

fusion
The nuclear process that combines light elements into heavier ones. Fusion is the principle behind the hydrogen bomb and the source of power of stars. Most stars, including the Sun, use the energy derived from the fusion of hydrogen into helium but some stars (e.g. red giants) can fuse helium into carbon.

galactic clusters
Groups of stars of the same age loosely bound together over a size of several light years. For this reason, they are also referred to as 'open clusters'. They are generally found in the disk or spiral aims of a galaxy. The most well-known example of a galactic cluster is the Pleiades (Fig. 9.1). Most of the galactic clusters are young, with ages of less than a few billion years.

galaxies
A large collection of stars grouped together by gravitational attraction. There are approximately 100 billion stars in our own Milky Way galaxy.

global clusters
In contrast to galactic clusters, globular clusters are tightly bound with many thousands, sometimes millions, of stars within a volume of 100 light years across. Globular clusters are generally located in the halos of galaxies and are believed to have formed over ten billion years ago.

graphite
A carbon material made up of rings of six carbon atoms arranged in horizontal layers. Although graphite lacks distinctive spectral signatures, it is believed to be produced in the winds of carbon stars.

helium
The second lightest element after hydrogen. Since it is lighter than air, helium is often used to fill balloons.

HST
The *Hubble Space Telescope*. Built at a cost of about US $3 billion (depending on how the cost is counted) and placed into Earth's orbit by the space shuttle in 1990, it is the premier astronomical instrument. It has been serviced three times by the shuttle crew in 1993, 1997 and 1999. The beautiful

astronomical images taken by the *HST* have generated a high degree of public interest in astronomy.

Hubble constant
A measure of the expanding rate of the Universe. A larger value of the Hubble constant implies a younger Universe.

infrared
The part of the electromagnetic spectrum with wavelengths longer than about 0.7 μm and shorter than about 100 μm. Although the short wavelength boundary is fairly well defined by the limit of human eye vision, there is no precise long-wavelength boundary between the far infrared and the submillimeter wave.

luminosity
The total power radiated by a star over all wavelengths. Two stars of the same luminosity but of different colors can have very different visual brightness. Although central stars of planetary nebulae are very luminous, they radiate primarily in the ultraviolet and can be quite faint in the visible.

interferometry
The technique used to create images of very high resolution by combining an array of telescopes to simulate a very large telescope.

ions
Atoms that have been stripped of one or more electrons.

Mach number
The ratio of object speed to the sound speed.

magnitude
A measure of stellar brightness in logarithmic scale. For curious reasons, the smaller the magnitude, the brighter the star. A star of 0th magnitude is 100 times brighter than a star of 5th magnitude. The unaided human eye can see stars down to about 6th magnitude.

main sequence
A band on the Herztsprung–Russell diagram on which most stars (including the Sun) are located. They represent the typical colors and brightness a star would have (depending on their mass) if they are burning hydrogen in the core. Red giants and central stars of planetary nebulae are evolved stars which have exhausted hydrogen in the core and are much brighter than most stars on the main sequence.

mass loss
Most stars eject matter from their surfaces in the form of stellar winds. Asymptotic giant branch stars in particular have very strong winds and very high mass loss rates.

micron or micrometer (μm)
10^{-6} meter.

molecule
A molecule is made up of two or more atoms. For example, the water molecule has two hydrogen and one oxygen atoms.

neutron star
A possible end state of evolution of massive stars after exploding as supernovae.

NGC* and *IC
The *New General Catalogue of Nebulae and Clusters of Stars* by J.L.E. Dreyer containing 7840 objects was first published in 1888. It was later supplemented by the *Index Catalogue* (*IC*) of 5386 objects. Most of the objects in the catalogue are galaxies, star clusters, and planetary nebulae.

Palomar Sky Survey
The *National Geographic Society–Palomar Observatory Sky Survey* was carried out by the 48-inch Schmidt telescope. It covered the sky from the north celestial pole to declination −27 on 879 pairs of plates, one blue-sensitive and one red-sensitive.

proton
An elementary particle which together with neutrons form the nuclei of atoms.

radio waves
Electromagnetic waves with wavelengths longer than approximately 10 cm. Radio waves with wavelengths shorter than 10 cm are sometimes referred to as microwaves, but there is no precise boundary between the two. Most radio and TV broadcasts use radio waves with wavelengths longer than 1 m. Modern cellular phones utilize radio waves of approximately 30 cm. Astronomical radio telescopes, such as the *VLA*, operate in the 1–100 cm range.

red giants
Red giants refer generally to stars that have evolved off the main sequence to higher luminosities and lower temperatures. They are separated into first giant branch stars and asymptotic giant branch stars based on the nuclear energy source. First giant branch stars have a helium core and burn hydrogen in a shell, whereas asymptotic giant branch stars have a carbon–oxygen core and burn hydrogen and helium in shells.

red shift
The shift of spectral lines of galaxies to longer wavelengths as the result of the expansion of the universe.

shock waves
Waves generated when an object is moving at a speed faster than sound. Shock waves are common in the interstellar medium because the sound speed is low due to the low density.

spectral type
The alphabetical sequence O, B, A, F, G, K, M is used to classify stars of decreasing temperatures. O stars have temperatures of approximately 40 000 °C and M stars have temperatures of approximately 3000 °C.

spectrum
The separation of light into different colors. Modern astronomical spectrographs are much more advanced than the prism first used by Newton and are capable of separating light into very fine wavelength intervals. Since atoms and molecules emit light at unique wavelengths, astronomical spectroscopy allows the detection of the presence of atoms and molecules and the measurements of the quantities present.

stars
Stars are self luminous, spherical gaseous objects that are held together by gravitational self attraction. The power of stars is provided by nuclear reactions in the center. Our nearest star is the Sun.

stellar wind
Streams of matter flowing from the surface of stars. Although most stars (including the Sun) have stellar winds, the magnitudes of the winds are strongest in red giants and O and B stars.

submillimeter wave
There is no precise definition of the submillimeter part of the electromagnetic spectrum, but it generally refers to wavelengths shorter than 1 mm and longer than 100 μm. The Earth's atmosphere is mostly opaque over the entire submillimeter-wave region and observations have to be carried out in high-flying aircraft or satellites.

supernovae
Powerful stellar explosions resulting in visual brightening of a billion times over the original stellar brightness. Supernovae occur at the end of the evolution of massive stars (over ten times the mass of the Sun), after all nuclear fuels have been exhausted. However, low mass stars in a binary system can also cause a supernova explosion if a white dwarf receives too much matter from its companion, causing its mass to exceed the Chandrasekhar limit.

ultraviolet
The part of the electromagnetic spectrum with wavelengths shorter than 0.3 μm and longer than 100 Å (1 Å = 10^{-8} cm).

theorists
Astronomers who apply the laws of physics and the techniques of mathematics and computer science to simulate the structure of stars, nebulae, galaxies, and the Universe.

wavelength
A quantitative measure of color. The human eye response range extends from red (corresponding to a wavelength of approximately 0.7 μm) to violet (a wavelength of approximately 0.3 μm).

Wien's law
The relation between the color and temperature of a radiating object. A red giant of a temperature of 3000 °C (e.g. Antares) will radiate primarily in the near infrared, whereas a star of 10 000 °C (e.g. Vega) will radiate primarily in the near ultraviolet.

white dwarfs
The end state of most stars. White dwarfs are small, dense, and faint.

X-ray
The part of the electromagnetic spectrum with wavelengths shorter than 100 Å and longer than ~0.1 Å.

21 Some commonly observed planetary nebulae

NGC	Common name	RA (J2000)			DEC(J2000)			Constellation
		h	m	s	°	'	"	
40		00	13	01.0	+72	31	20	Cepheus
650-1	Little Dumbbell	01	42	24	+51	34		Perseus
1514		04	09	17.1	+30	46	35	Tarus
IC 418		05	27	28.3	−12	41	48	Lepus
2346		07	09	22.1	−00	48	17	Monoceros
2371-72		07	25	35.3	+29	29	36	Gemini
2392	Eskimo	07	29	10.7	+20	54	37	Gemini
2440		07	41	55.4	−18	12	33	Puppis
2899		09	27	03.5	−56	06	18	Velorum
3132	Southern Ring	10	07	01.8	−40	26	10	Velorum
3242		10	24	45.9	−18	38	38	Hydra
3587	Owl	11	14	46.5	+55	01	00	Ursae Major
6302		17	13	44.3	−37	06	06	Scorpius
6309		17	14	03.6	−12	54	37	Ophiuchus
6369		17	29	20.8	−23	45	32	Ophiuchus
6543	Cat's eye	17	58	33.4	+66	37	59	Draco
6720	Ring	18	53	53.7	+33	01	40	Lyra
6772		19	14	28.4	−02	42	26	Aquila
6781		19	18	28.3	+06	32	23	Aquila
BD+30 3639		19	34	45.2	+30	30	59	Cygnus
6826		19	44	48.3	+50	31	30	Cygnus
6853	Dumbbell	19	59	36.2	+22	43	01	Vulpecula
6894		20	16	12.9	+30	33	54	Cygnus
7009	Saturn	21	04	10.8	−11	21	57	Aquarius
7026		21	06	18.5	+47	51	07	Cygnus
7027		21	07	01.7	+42	14	10	Cygnus
7293	Helix	22	29	38.7	−20	50	15	Aquarius
7662		23	25	53.8	+42	32	06	Andromeda

22 Further reading

Josif S. Skhlovsky (1916–1985) was responsible for the fundamental understanding of the origin of planetary nebulae. His book, *Stars their Birth, Life, and Death*, was translated into English in 1978 (W.H. Freeman & Co: San Francisco). The book contains an excellent summary of stellar structure and evolution, and gives us a unique historical perspective. Lawrence H. Aller is widely considered as the father of modern planetary nebulae research. Through the application of quantum physics, he was able to interpret the spectra of planetary nebulae and derived the physical properties of the nebulae. His book *Atoms, Stars, and Nebulae* (3rd edition, Cambridge University Press) is a good introduction to this field for students with a university level physics background. A more specialized and advanced treatment can be found in his 1956 book *Gaseous Nebulae* (Chapman and Hall: London). The book *Planetary Nebulae* by Stuart Pottasch in 1984 (Reidel: Dordrecht) contains an excellent overview of the theory and observations of planetary nebulae. A more recent review, emphasizing results of space observations and current theoretical developments, can be found in the book *The Origin and Evolution of Planetary Nebulae* by Sun Kwok (2000, Cambridge University Press).

The first popular account of planetary nebulae was given by Lawrence Aller through a series of 14 articles in *Sky and Telescope* magazine and later collected into a monograph published by Sky Publishing Corporation in 1971. The recent surge in interest in planetary nebulae has resulted in many popular articles on this subject. A partial list includes 'Recent Findings about Planetary Nebulae' by Yervant Terzian (*Sky and Telescope*, December 1977), 'Not with a Bang but a Whimper' by Sun Kwok (*Sky and Telescope*, May 1982), 'Planetary Nebulae' by Noam Soker (*Scientific American*, May 1992), 'A Modern View of Planetary Nebulae' by Sun Kwok (*Sky and Telescope*, July 1996), 'When Planetaries meet Planets' by Arsen Hajian (*Mercury*, May 1996), 'The Shapes of Planetary Nebulae' by Bruce Balick (*American Scientist*, July 1996), 'Stellar Metamorphosis' by Sun Kwok (*Sky and Telescope*, October 1998), 'Bursting the Bubbles' by Adam Frank (*Astronomy*, April 2000), and 'What is the True Shape of the Ring Nebula?' by Sun Kwok (*Sky and Telescope*, July 2000).

For technical readers who are interested in tracing the development of the field of planetary nebulae, the best reference material is the proceedings of the symposia organized by the International Astronomical Union. This series began with IAU Symp. 34 in Tatranska Lomnica, Czechoslovakia in 1967 (*Planetary Nebulae*, eds. D.E. Osterbrock and C.R. O'Dell, Reidel, 1968), followed by IAU Symp. 76 in Ithaca, NY, USA, in 1976 (*Planetary Nebulae*, ed. Y. Terzian, Reidel, 1978), IAU Symp. 103 in London, England in 1982 (*Planetary Nebulae*, ed. D.R. Flower, Reidel, 1983), IAU Symp. 131 in Mexico City, Mexico in 1987 (*Planetary Nebulae*, ed. S. Torres-Peimbert, Kluwer, 1989), IAU Symp. 155 in Innsbruck, Austria in 1992 (*Planetary Nebulae*, eds. R. Weinberger and A. Acker, Kluwer, 1993), and IAU Symp. 180 in Groningen, Holland in 1996 (*Planetary Nebulae*, eds. H.J. Habing and H.J.G.L.M. Lamers, Kluwer 1997). The next IAU symposium is scheduled to take place in Canberra, Australia, in November 2001.

23 Notes on images

Unlike stars, which emit continuous colors, planetary nebulae emit most of their visible light in atomic line radiation. It is therefore much more efficient to image planetary nebulae through narrow filters centered on the wavelengths of various atomic lines. Most of the color images of planetary nebulae in this book are color composites of monochromatic pictures taken with these filters. Each of these monochromatic images is assigned a color, e.g. red for singly ionized nitrogen, green for the Hα transition of hydrogen, and blue for doubly ionized oxygen. Because these color assignments are arbitrary, the colors as shown in these pictures do not resemble the colors we perceive with human eyes. So it is possible for a planetary nebula to be shown in different color renditions, because a different filter combination or color assignment is used. Since these narrow band observations do not represent the continuous colors of stars, the background stars in some of the planetary nebulae images do not appear white as they should.

The *HST* Wide Field Planetary Camera 2 (WFPC2) instrument has a high dynamic range and can detect very faint nebulosities. In order to display the low-level emissions, we have sometimes artificially compressed the intensity scale in the picture. Since most planetary nebulae have strong emissions in doubly ionized oxygen or hydrogen (Hα), their true colors are mostly green or red. However, faint emissions are often only detected in the singly ionized nitrogen filter and we have sometimes exaggerated the contribution of this filter in the color composite picture. While these pictures serve a very useful purpose of delineating the complex structures of planetary nebula, they are not suitable for the derivation of physical parameters by quantitative analysis.

Proto-planetary nebulae do not radiate line radiation, and their colors are due to the reflection of light from their central stars. The color images of proto-planetary nebulae in this book are therefore mostly constructed from color composites of images taken with broad filters centered on blue, yellow, and red and their color appearances are closer to the color perception of the human eye. Since proto-planetary nebulae often contain dust, which preferentially absorbs blue color over red color, the red parts of the nebulae are likely to have a higher concentration of dust.

Because the human eye is not particularly good at discriminating color (responding mostly to yellow), if we insist on reproducing color pictures of astronomical objects as they would be seen by human eyes, we would miss out a lot. The pictures in this book represent what you may see if you have a broader color perception, a wider range of brightness reception, as well as better sensitivity for faint light.

Color is sometimes used to represent a brightness scale in a monochromatic image. This is known as a false-color picture. Several false-color pictures are used in this book, either because we wish to emphasize the nebular image at one particular filter or because observations only in one filter are available.

The WFPC2 instrument consists of four cameras: a Planetary Camera (PC) and three Wide Field Cameras (WFC). Most of the planetary nebulae shown in this book are small enough that they fit in the field of view of the PC, but some larger images are made with a mosaic of the images by all four cameras.

Most of the *HST* images in this book were processed from archived data in the Canadian Astronomy Data Center. These data were calibrated using the latest recalibration software and reference files. The National Optical Astronomy Observatory Image Reduction and Analysis Facility and the Space Telescope Science Data Analysis System were used to remove the cosmic rays and to combine multiple exposures. Images that were obtained elsewhere are credited to the individual sources in the figure captions.

Further details on the color images are given below. Information includes the telescope and instrument used, the color assignment and the filters used, and the date of observation if available. For *HST* observations, the original principal investigator who carried out the observations is also listed.

CHAPTER 1

M27 and M57: 0.4-m telescope of KPNO, broad band blue, broad band green, broad band red, June 15, 2000, Adam Block.
NGC 2392: *HST* WFPC2, red: NII, green: Hα, blue: Hα, November 16 and December 1, 1998, PI: Borkowsky.
NGC 7293: AAT, photographic plates, D. Malin.
NGC 6369: *HST* WFPC2, red: I (f814W), green: average of I and V, blue: V (f555W), August 9, 1995, PI: H. Bond.
NGC 3132: *HST* WFPC2, red: NII, green: Hα, blue: OIII, observed on December 7, 1995, PI: J. Trauger.
IC 418: *HST* WFPC2, red: NII, green: Hα, blue: OIII, February and September 1999.

CHAPTER 2

NGC 6891: *HST* WFPC2, red: I, green: average of I and V, blue: V, November 20, 1995, PI: H. Bond.
IC 2165: *HST* WFPC2, red: Strömgren *y*, green: Strömgren *v*, blue: Strömgren *u*, November 26, 1995, PI: Westphal.
NGC 6302: *VLT*, broadband red, yellow and blue, May 1998.
M2-9:*HST* WFPC2, red: SII, green: NII, blue: Hα, August 7, 1997, PI: B. Balick.
MyCn 18: *HST* WFPC2, red: NII, green: OI, blue: OIII, July 30, 1995, PI: J. Trauger.
Shapely 1, AAT, photographic plates, D. Malin.
Wray 17-31: AAT, photographic plates, D. Malin.
NGC 6543, *HST* WFPC2, red: OI, green: HeI, blue: OIII, September 18, 1994, PI: P. Harrington.

CHAPTER 3

NGC 2440: *HST* WFPC2, red: I, green: average of V and I, blue: V, November 18, 1995, PI: Bond.

CHAPTER 4

A30: red: Hα, green: OIII, blue: OIII, C.T. Hua.
A39: red: NII, green: Hα, blue: OIII, C.T. Hua.
M1-61, M2-43, M3-35, IC 5117: VLA λ2cm, Aaquist and Kwok.
He2-320: *HST* WFPC2, red: NII, green: Hα, blue: Hα, August 23, 1999, PI: R. Sahai.
He2-180: *HST* WFPC2, red: NII, green: Hα, blue: Hα, August 12, 1999, PI: R. Sahai.
M1-73: *HST* WFPC2, red: NII, green: Hα, blue: Hα, July 15, 1999, PI: R. Sahai.
M1-12: *HST* WFPC2, red: Hα, green: NII, blue: NII, October 8, 1999, PI: R. Sahai.
He2-86 *HST* WFPC2, red: NII, green: Hα, blue: Hα: September 24, 1999: PI: R. Sahai.

CHAPTER 9

NGC 7027: *HST* WFPC2, V and I, August 21, 1995, PI: H. Bond.

CHAPTER 10

NGC 6751: *HST* WFPC2, red: NII, green: Hα, blue: OIII, April 21, 1998, PI: Hajian.
NGC 6828: KPNO, grey scale in Hα, G. Jacoby.
NGC 6543, KPNO, grey scale in Hα, G. Jacoby.
He2-119: grey scale in NII, C.T. Hua.
NGC 6891: Nordic Optical Telescope, grey scale in NII, Luis Miranda and Martin Guerrero.
NGC 3568: *HST* WFPC2, red: I, green: average of I and V, blue: V, August 4, 1995, PI: H. Bond.
NGC 5882: *HST* WFPC2, red: I, green: average of I and V, blue: V, July 25, 1995, PI: H. Bond.
NGC 5979: *HST* WFPC2, red: I, green: average of I and V, blue: V, February 7, 1997, PI: H. Bond.
NGC 6578: *HST* WFPC2, red: I, green: average of I and V, blue: V, August 16, 1995, PI: H. Bond.

NGC 2792: *HST* WFPC2, red: I, green: average of I and V, blue: V, August 11, 1995, PI: H. Bond.
IC 2448: *HST* WFPC2, red: I, green: average of I and V, blue: V, October 10, 1995, PI: H. Bond.
NGC 6629: *HST* WFPC2, red: I, green: average of I and V, blue: V, August 16, 1995, PI: H. Bond.
NGC 2022: *HST* WFPC2, red: I, green: average of I and V, blue: V, August 11, 1995, PI: H. Bond.
NGC 6543: *Chandra*, blue: X-ray.

CHAPTER 11

NGC 6309: *HST* WFPC2, red: I, green: average of I and V, blue: V, August 26, 1995, PI: H. Bond.
NGC 7026: *HST* WFPC2, red: NII, green: V, blue: OIII, July 7, 1998, PI: Hajian.
NGC 2346: *HST* WFPC2, red: NII, green: Hα, blue: OIII, March 6, 1997, PI: Chapman.
NGC 6302: red: NII, green: Hα, blue: OIII, C.T. Hua.
Hubble 5: *HST* WFPC2, red: NII, green: Hα, blue: OIII. September 9, 1997, PI: B. Balick.
NGC 6537: *HST* WFPC2, red: NII, green: Hα, blue: OIII, September 12, 1998, PI:Balick.
NGC 6881: *HST* WFPC2, red: NII, green: Hα, blue: OIII, November 1, 1999, PI: Kwok.
NGC 6790: *HST* WFPC2, red: NII, green: Hα, blue: OIII, October 31, 1999, PI: Kwok.
Hb 12: *HST* WFPC2, false color Hα, April 6, 1996, PI: Sahai.
He2-104: *HST* WFPC2, false color Hα, May 22, 1999, PI: Corradi.
NGC 6720: *HST* WFPC2, red: NII, green: OIII, blue: HeI, October 16, 1998, PI: H. Bond.
Sh1-89: grey scale NII, C.T. Hua.
SaWe3:grey scale NII, C.T. Hua.
NGC 6720: Subaru, false color in Hα.

CHAPTER 12

NGC 6720: False color 2.12 μm, D. Thompson.
NGC 6781: 2MASS, red: 2.2 μm, green: 1.6 μm, blue: 1.2 μm.

NGC 3918: *HST* WFPC2, red: I, green: average of I and V, blue: V, October 18, 1995, PI: H. Bond.

CHAPTER 13

Egg Nebula: *HST* WFPC2, false color wide V (f606W), April 9, 1998, PI: Sahai.
M1-61: *HST* WFPC2, red: NII, green: Hα, blue: OIII, November 8, 1999, PI: Kwok.

CHAPTER 14

Cotton Candy Nebula: *HST* WFPC2, red: I, green: wide V, blue: wide B (f450W), May 14, 1997, PI: Kwok; October 22, 1996, PI: Bobrowsky.
Silkworm Nebula: *HST* WFPC2, red: I, green: wide V, blue: wide B, May 22, 1997, PI: Kwok; April 26, 1996, PI: Bobrowsky.
Water Lily Nebula: *HST* WFPC2, red: I, green: average of I and wide V, red: wide V, March 1997, PI: S. Kwok, June 28, 1999, PI: B. Hrivnak.
Walnut Nebula: *HST* WFPC2, red: I, green: average of I and wide V, blue: wide V, March 1997, PI: S. Kwok; June 2, 1999, PI: B. Hrivnak.
Spindle Nebula: *HST* WFPC2, red: I, green: average of I and wide V, blue: wide V, May 30, 1999, PI: B. Hrivnak.
Boomerang Nebula: *HST*WFPC2, grey scale wide V, March 23, 1998, PI: Trauger.
M1-92: *HST* WFPC2, red SII, green NII, blue Hα, May 30, 1996, PI: Bujarrabal.
Hen 3-401: *HST* WFPC2, red: SII, green Hα, blue: wide V, June 12, 1997, PI: Sahai.
Roberts 22: *HST* WFPC2, red SII, green Hα, blue: wide V, April 11-12, 1997, PI: Sahai.
Mz 3: *HST* WFPC2, red: NII, green: Hα, blue: HB, June 30, 1998, PI: Trauger.
IC 4406: *VLT*, broad band red, yellow and blue.
Trapezium: *HST NICMOS*, red: H, green: average of J and H, blue: J.
Egg Nebula: *HST* WFPC2 and *NICMOS*, red: H_2, green: H, blue: wide V, April 9, 1998, PI: Sahai.

NGC 7027: WFPC2 and *NICMOS*: red H_2, green: 1.9 μm, blue: average of V and I, October 9, 1997, PI: Latter.

Cotton Candy: *HST* WFPC2 and *NICMOS*: red: H_2, green: I, blue: wide V, August 16, 1998, PI: Kwok.

CHAPTER 15

NGC 6826: *HST* WFPC2, red: SII, green: Hα, blue: NII, January 27, 1996, PI: Balick.

NGC 7009: *HST* WFPC2, red: SII, green: Hα, blue: OI, April 28, 1996, PI: B. Balick.

NGC 3242: *HST* WFPC2, red: SII, green Hα, blue: OIII, April 19, 1996, PI: B. Balick.

NGC 7662: *HST* WFPC2, red: NII, green Hα, red: OIII, January 5, 1996, PI: B. Balick.

IC 4593: *HST* WFPC2, red: NII, green: Hα, blue: Hα, March 31 and May 6, 1999, PI: Borkowsky.

NGC 2392: *HST* WFPC2, red: NII, green: Hα, blue: OIII, violet: HeI, January 10-11, 2000, PI: Fructher.

NGC 2392: *HST* WFPC2, grey scale Hα, November 16, 1998, PI: Borkowsky.

NGC 7293: *HST* WFPC2, red: NII, green: Hα, blue: OIII, November2, 1995, PI: O'Dell.

IC 4663: *HST* WFPC2, red: NII, green: V, blue: OIII, August 4, 1998, PI: Hajian.

NGC 6543: *HST* WFPC2, grey scale NII, September 18, 1994, PI: Harrington.

NGC 7354: *HST* WFPC2, red: NII, green:V, blue: OIII, July 21, 1998, PI: Hajian.

Hen2-447: *HST* WFPC2, red: NII, green: Hα, blue: OIII, November 9, 1999, PI: Kwok.

NGC 2440: *HST* WFPC2, red: I, green: average of I and V, blue: V, November 18, 1995, PI: Bond.

K3-35: *VLA*, λ2cm, Aaquist and Kwok.

NGC 5307: *HST* WFPC2, red: I, green: average of I and V, blue: V, July 28, 1995, PI: H. Bond.

He3-1475: *HST* WFPC2, red: I, green: NII, blue: V, June 23 and 26, 1996, PI: Bobrowsky.

IC 4634: *HST* WFPC2, red: NII, green: Hα, blue: Hβ, May 15, 1998, PI: J. Trauger.

NGC 6884: *HST* WFPC2, red: I, green: average of I and V, blue: V, October 13, 1995, PI: Bond.

NGC 6543: *HST* WFPC2, grey scale OIII, September 18, 1994, PI: Harrington.

CHAPTER 16

M15: *HST* WFPC2, red: I, magenta: Hα, green: Strömgren *y* and OIII, blue: B, December 15, 1998, PI: Bond.

CHAPTER 18

IRAS 04296+3429: *HST* WFPC2, red: R, green: V, blue: B, August 2, 1997, PI: Bobrowsky; January 15, 1999, PI: Trammell.

CHAPTER 19

BD+30+3639: *HST* WFPC2, red: SIII, green: SII, blue: NII, March 6, 1994, PI: Harrington.

IRAS 21282+5050: *HST* WFPC2, false color in Hα, August 23, 1999, PI: Sahai.

Index